Bilingual Public Schooling in the United States

Bilingual Public Schooling in the United States

A History of America's "Polyglot Boardinghouse"

Paul J. Ramsey

BILINGUAL PUBLIC SCHOOLING IN THE UNITED STATES
Copyright © Paul J. Ramsey

First published in 2010 by
PALGRAVE MACMILLAN®
in the United States—a division of St. Martin's Press LLC,
175 Fifth Avenue, New York, NY 10010.

Where this book is distributed in the UK, Europe and the rest of the world,
this is by Palgrave Macmillan, a division of Macmillan Publishers Limited,
registered in England, company number 785998, of Houndmills,
Basingstoke, Hampshire RG21 6XS.

Palgrave Macmillan is the global academic imprint of the above companies
and has companies and representatives throughout the world.

Palgrave® and Macmillan® are registered trademarks in the United States,
the United Kingdom, Europe and other countries.

ISBN: 978–0–230–61851–0

Library of Congress Cataloging-in-Publication Data

Ramsey, Paul J.
    Bilingual public schooling in the United States : a history of
America's "polyglot boardinghouse" / Paul J. Ramsey.
        p. cm.
    Includes bibliographical references and index.
    ISBN 978–0–230–61851–0 (alk. paper)
        1. Education, Bilingual—United States—History. 2. Multicultural
education—United States—History. 3. Minorities—Education—
United States—History. 4. Language and education—United States—
History. 5. Education and state—United States. I. Title.

LC3731R285 2010
370.117'50973—dc22   2009039970

A catalogue record of the book is available from the British Library.

Design by Newgen Imaging Systems (P) Ltd., Chennai, India.

First edition: March 2010

10 9 8 7 6 5 4 3 2 1

Printed in the United States of America.

For Ronald Takaki (1939–2009), a remarkable scholar and an immense inspiration.

Contents

Tables

Acknowledgments

No historical work is a solitary endeavor, although, while sitting alone in a library, in an archive, or in front of a computer screen, it sometimes may feel as though it is. Despite the occasional feelings of solitude, historians recognize the contributions that others make to their work, and, therefore, I would like to thank those who, in a variety of ways, helped me with this somewhat ambitious project. Although I believe that this book is unique and fills a crucial void in the historical literature, it, because of its breadth, could not have been accomplished without standing on the shoulders of giants. Path-breaking researchers, some of whom I have never met or have long since passed, assisted me—without their knowledge—through their scholarship. Among those who aided this project most are Theodore Andersson, Carlos Kevin Blanton, John Bodnar, John Higham, Carl F. Kaestle, Walter D. Kamphoefner, Michael B. Katz, David M. Kennedy, John M. Nieto-Phillips, William J. Reese, George I. Sanchez, Steven L. Schlossman, Ronald Takaki, David B. Tyack, Robert H. Wiebe, and—although this study is critical of his work—Heinz Kloss.

This book emanates from my dissertation at Indiana University, and, in some respects, the topic came to me by happenstance. As a doctoral student, I had been doing a fair bit of research on the Indianapolis schools, particularly focusing on the influence of the powerful German element over the educational affairs of the city. Simultaneously, Kate Rousmaniere was recruiting authors for the education section of *The American Midwest: An Interpretive Encyclopedia*. Kate contacted my academic advisor, Andrea Walton, and asked if she had any doctoral students willing to write an essay for the encyclopedia. Because of my interest in the nineteenth-century German-English program in Indianapolis, I agreed to write a short piece on bilingual education, a task that forced me to read more broadly in the dual-language literature. I was surprised to find that there were very few comprehensive studies of the history of bilingual education and that this handful of histories had, in my view, serious shortcomings. This enormous gap in the historiography essentially defined the parameters of my dissertation. I cannot thank Kate enough for unwittingly helping me find my topic, nor can I fully express my thanks to my dissertation committee members—Andrea, B. Edward McClellan, Donald Warren, and Barry L. Bull—for allowing me a great deal of latitude for this slightly overzealous project.

Although Andrea, Ed, Don, and Barry handed me a "blank check"—as Barry called it during my proposal defense—for the contours of my study, they gave me a great deal of guidance and support during my project, for which I am sincerely grateful. Like the scholars mentioned before, I stood on my committee members' shoulders, but for a longer period of time. These scholars are truly intellectual giants, and I find it difficult to articulate the influence their research, courses, and conversations have had on me and my work; I am eternally indebted. The work and character of Andrea, Ed, Don, and Barry have been an enormous inspiration for my academic research. Certainly, others have inspired me as well. My undergraduate history professors, most notably Donald E. Pitzer, Casey Harison, and the late William McGucken, cultivated my love of history, while my German professors, the late Donald Wolfe and Susan Smith Wolfe, piqued my interest in languages. At the graduate level, Lawrence J. Friedman and Michael Grossberg nurtured my historical thinking. Nevertheless, my deepest gratitude is reserved for Andrea, Ed, Don, and Barry.

This project could not have been accomplished without the assistance of numerous librarians and archivists, primarily those at the Indiana University (IU) and the University of Illinois libraries and the Nettie Lee Benson Latin American Collection at the University of Texas, Austin. These diligent folks helped me locate invaluable sources for this study. The dissertation also would not have been possible without the generous financial support of the Indiana University School of Education, the Spencer Foundation, and the Office of the Provost at Eastern Michigan University (EMU). I received several travel and research grants from the IU School of Education and its Department of Educational Leadership and Policy Studies in order to explore libraries and archives; the Spencer Foundation's Dissertation Fellowship the EMU Provost's Research Award for New Faculty supported much of the writing phase of this project.

There are numerous others who also contributed to this project. During my studies, a handful of doctoral students and I would get together weekly to discuss classic and contemporary historical and philosophical works, as well as our own scholarship. At these meetings—frequently held at a cafeteria with an artsy ambiance—Joshua B. Garrison and Glenn Lauzon, incredibly knowledgeable and talented historians of education, provided me with a great deal of insight in the early stages of my project. As the historians began to graduate, the weekly get-togethers morphed into a philosophy group, which met on Monday nights at a local pub. Although I called it the "Goodbye Blue Monday" Reading and Drinking Society, the name never stuck because, other than me, no members were (obsessive) Kurt Vonnegut fans and, more importantly, Thursday evenings became the preferred meeting time. Nevertheless, Chris Hanks and Dini Metro-Roland, both gifted philosophers of education, read parts of this dissertation at our meetings and gave me valuable feedback. I am grateful for the assistance and support my fellow doctoral students gave me while I was working on this project.

I would also like to thank the staff, anonymous reviewers, and editors at Palgrave Macmillan, especially Julia Cohen, Burke Gerstenschlager, Samantha Hasey, and Allison McElgunn, for providing thoughtful comments regarding this project. My colleagues and students at Eastern Michigan University have been exceptionally supportive of this book, most notably Joe Bishop, who read early drafts of one of the chapters, and Serhiy Kovalchuk, who assisted with the index. In addition, I turned much of the third chapter of this book—along with bits from the introduction and chapter 1—into a free-standing essay, titled "In the Region of Babel: Public Bilingual Schooling in the Midwest, 1840s–1880s." The essay won the History of Education Society's prestigious Henry Barnard Prize in 2007 and was published in the *History of Education Quarterly* in 2009. Many thanks, therefore, go out to the Barnard Prize Committee and the editors and staff of *HEQ*.

Lastly, I would like to thank my wife and children for putting up with my nearly obsessive behavior during the duration of this project. They patiently and lovingly supported me while I was researching and writing the dissertation and book. Unfortunately, now I do not have a readymade excuse for my neuroses.

Introduction

Drafting the Blueprints for This Old Boardinghouse

In the midst of World War I, Indiana University's James Woodburn addressed the Indiana State Teachers Association. "Let us strive," the professor of history told the Hoosier teachers, "to save America from being a polyglot nation—a conglomeration of tongues and nationalities, like a 'polyglot boardinghouse,' as Mr. [Theodore] Roosevelt has put it."[1] Woodburn's fear of a multilingual society paralleled the attitudes of many political leaders, scholars, and educators during and after the Great War, attitudes that greatly altered the nature of bilingual instruction throughout the United States. While Woodburn and others' desire for a *united nation*—with...*one* language" was largely an outgrowth of the anxiety that the international conflict had created, the fear of an America that served as a "polyglot boardinghouse" had deeper roots, and that fear—although ebbing and flowing—has lingered into the present, particularly with regard to bilingual education in America's public schools.[2]

Of course, Roosevelt's use of the phrase "polyglot boardinghouse" was intended to inspire fear, but the notion of a multilingual America need not necessarily hold a negative connotation; the response to this notion is often a matter of perspective. Senator Listor Rosewater, a character in a Kurt Vonnegut novel, underscored these diverse perspectives regarding a multiethnic society. Discussing the decline of ancient Rome, the conservative senator stated that the leftists "said what liberals always say after they have led a great nation to such a lawless, self-indulgent, polyglot condition: 'Things have never been better! Look at all the freedom! Look at all the equality!'" The liberals, Rosewater continued, "loved the barbarians so much they wanted to open all the gates, have all the soldiers lay their weapons down, and let the barbarians come in!"[3] While the senator feared a multilingual, pluralistic society, others, as Rosewater suggested, welcomed such changes, and that difference of opinion—although Senator Rosewater's view has traditionally dominated the conversation—has long been a part of America's past.

The United States has struggled with linguistic diversity throughout its history. Even before the nation's founding, prominent Anglos sensed and often fueled English-speakers' uneasiness with the growing linguistic pluralism in the colonies. Writing about his beloved Pennsylvania, Benjamin Franklin noted in 1753 that "unless the stream of" immigration could be quelled, the Germans "will soon so out number us, that...we will not, in My Opinion, be able to preserve our language, and even our government will become precarious."[4] By the nineteenth century, when immigration from non-English-speaking nations dramatically increased, the fear of a polyglot America became, at various times, a particularly salient concern for many in the nation. It was, after all, during this century that the German-language myth— the myth that suggested the United States, in its founding years, was one vote shy of becoming a German-speaking nation—took root. Although simply a legend—the actual incident concerned the publication of official documents in German for the immigrants, not a switch in the national language—many German Americans were, nonetheless, fond of this bit of nineteenth-century lore.[5]

The coming of the common schools in the second quarter of the nineteenth century added new layers to the uneasiness over language diversity. Those early public schools, intensely local in their governance, often became institutions for maintaining the linguistic and cultural heritages of ethnic communities, particularly in those towns and sections of the ever-growing cities where immigrants controlled their own affairs. Bilingual education, therefore, became a staple of the common school experience in many ethnic enclaves, but it also sparked controversy and, at times, hostility. John B. Peaslee, for one, admitted that he was "prejudiced against teaching children any foreign language" when he first became a teacher in the German-English public schools of nineteenth-century Cincinnati; his "prejudice" quickly faded as he witnessed the remarkable success the city's schools had with educating immigrant children.[6] Unlike Peaslee, many onlookers were not convinced of the value of bilingual instruction in America's schools, and the opposition to the polyglot boardinghouse only intensified as wave after wave of immigrants washed onto American shores during the decades surrounding the turn of the twentieth century.

These opponents, many of whom were nativists of various degrees, found in England the language and culture that should be the primary foundation of the United States. From their perspective, the diversity brought forth by the immigrants—particularly the politically powerful groups, such as the Germans—was essentially a declaration of "war against Anglo-Saxonism" in America.[7]

The fear of a polyglot America—a multilingual America that is partially sustained through dual-language instruction—is still far from being laid to rest. Even after the tumultuous post-WWII era brought forth calls for racial and ethnic equality (and the period witnessed several scattered accomplishments), many Americans today still feel uncomfortable with linguistic diversity, especially in the public schools. The current debate over immigration policies—a debate that includes not only naturalization, security, and economic issues, but also the place of English in American society—attests to this continued uneasiness.[8] Of course, bilingual education is a controversial educational issue not only because of a fear of the "polyglot boardinghouse," but also because of ongoing specialist debates over the nature of language acquisition, as well as a whole range of pedagogical and policy concerns. For example, Margaret Garcia Dugan, Arizona's deputy superintendent of public instruction, enforced the state's Proposition 203, the 2000 educational policy that limited bilingual education in the public schools. Dugan was an enthusiastic supporter of the policy because she intuitively believed that "most of the studies [on bilingual education] are invalid" and, therefore, advocated a monolingual approach throughout the state.[9]

As important as the language-acquisition debates are, it is the frightening image of a multilingual America that alarms many opponents of dual-language instruction and fuels campaigns for an English-only society. William Bennett, former secretary of education, is dismissive of bilingual instruction, noting that "[t]o be a citizen is to share in something common—in common principles, common memories, and a common language in which to discuss our common affairs. Our common language is, of course, English." Diane Ravitch, a former official in the U.S. Department of Education, also questions the idea of bilingual education because, in her view, its "aim is to use the public schools to promote the maintenance of distinct ethnic communities."[10] Of course, Ravitch is partially correct about the goal of bilingual instruction, and that potential "aim"—the perpetuation of "distinct ethnic communities"—is what inspires the anxiety over American pluralism.

Amid all of this tension surrounding bilingual education and its apparent goal of creating a polyglot boardinghouse, a historical study of the issues related to the controversy, surprisingly, has not yet emerged. A systematic history of public bilingual education's early years is also needed because many still do not realize that bilingual instruction has had a long and sometimes checkered past. This monograph attempts, at least partially, to correct some of these misunderstandings. The study explores the history of bilingual education in the public elementary schools in the United States (and in some of that nation's territorial holdings) from the 1840s to 1970s, from Cincinnati's adoption of German-English instruction to the federal government's involvement with bilingual schooling. The book examines the ways in which the larger society—its intellectual, cultural, demographic, economic, and moral currents—shaped the contours of bilingual education in America. This study also explores the power dynamics of bilingual education, dynamics that often shifted based on the larger context. That is, those who implemented and controlled bilingual programs in America's public schools were a diverse lot, ranging from ethnic leaders to nativists hoping to "Americanize" immigrant, colonized, and indigenous children. Naturally, power over these programs changed hands over time. Schooling, whether bilingual or not, imposed a set of values on foreign-language-speaking children, but they were not always passive recipients of America's public mores. Educational imposition frequently mixed with cultural traditions to create unintended outcomes within public schools and within ethnic communities.

The Drafters of the Blueprints: Historians, Linguists, and Educationists

In his 2004 study, *The Strange Career of Bilingual Education in Texas, 1836–1981*, Carlos Kevin Blanton noted that "historians have yet to document the story of American bilingual education," a point that historian Steven Schlossman made twenty years earlier.[11] Historians, of course, have not entirely neglected bilingual education in America's past, but the issue has not received the sustained historical attention that one would expect from such a contentious educational practice. There have been several excellent studies of the dual-language programs of the twentieth century, but, for the most part, historians have paid only passing attention to its earlier history.[12] Several prominent historians have touched upon bilingual schooling in their work, although the treatment of the topic is often of secondary importance, simply a supporting detail that demonstrates the immigrant experience in America.[13] Even historians who emphasize the education of American immigrants have devoted very little attention to the bilingual institutions where many immigrant children were educated.[14] While bilingual education has largely been on the periphery of historical inquiry, other scholars, particularly linguists, have not ignored the issue. Nevertheless, what has emerged from the scholarship of historians and linguists is a blueprint of the polyglot boardinghouse that has many shortcomings.

Resisting the all-encompassing concept of "American character" put forth by the "consensus" historians of the 1950s and 1960s, many contemporary historians who write on immigrants in the United States often focus on a single ethnic group. This literature does, occasionally, examine the bilingual educational practices of the various ethnic groups.[15] One shortcoming of this emphasis on a single ethnic group is that it potentially leads to history that uncritically chronicles the triumphs of a particular group of people.[16] In addition to the potential for celebratory history, a focus on a single ethnic group ignores the larger interactions and patterns among other immigrant, indigenous, and colonized groups; attitudes directed at one ethnicity—such as nativism—often have had an impact on other peoples, and solely emphasizing the experience of one group potentially misses those connections.

While many historians are engaged in examining a particular ethnic group, others, occasionally the same scholars, focus on a single location, which can lead to similar shortcomings. Yet, when exploring bilingual education, there is some justification for a local focus. Schooling, it should not be forgotten, was largely a local endeavor for much of America's past. In his study of two German neighborhoods in Philadelphia, historian Russell A. Kazal has justified micro-studies by noting that "[w]hen looking for answers to large questions, it sometimes helps to dig in small places."[17] However, the sole focus on a particular community, like the emphasis on a single ethnic group, has the potential of obscuring the larger patterns of bilingual schooling. Whatever the flaws, these local case studies represent historians' greatest contribution to the development of a history of bilingual education in America.

While historians of various sorts have, marginally, added to our understanding of bilingualism in American schools, it has been the linguists—particularly sociolinguists—and other scholars of foreign languages who have taken the lead in writing the history of bilingual education in the United States. In the 1950s, for example, Einer Haugen at the University of Wisconsin conducted his path-breaking research on the bilingual activity of early Norwegian Americans, while, two decades later, Shirley Brice Heath, a professor of linguistics at Stanford, explored the language policies of colonial America and the early republic. Theodore Andersson of the University of Texas and Joshua Fishman of Yeshiva University—both of whom gave expert testimony before Congress in the 1960s when it was considering the Bilingual Education Act—also contributed to the writing of the history of bilingual education in the United States.[18] Yet, before all these scholars, there was Heinz Kloss. During the Third Reich, Kloss worked as a sociolinguist in the American section of the German Foreign Institute (DAI), where he studied the Germans in the United States. Kloss's career continued well beyond the Nazi era, during which he wrote numerous scholarly works on the history of foreign-language speakers in America. Kloss's prolific scholarship earned him the reputation as one of the leading experts on bilingual education in the United States, but, recently, his work has been seriously criticized within the linguistic community by suggesting that even his later work carried the stain of Nazi ideology. With all of its problems, Kloss's scholarship, particularly his 1977 *American Bilingual Tradition*, remains, sadly, as one of the few surveys of the history of bilingual education in the United States. (This serious historical void is perhaps one of the reasons why Kloss's *Tradition* was republished in 1998.)[19]

This rough blueprint of the history of bilingual education left to us by linguists, particularly Kloss, is what education scholars draw upon when writing about the topic. Good scholars recognize that contemporary issues are rooted in a historical context, and when educationists attempt to steep their research in that context, they primarily have only this "poor history," as Blanton has called it, with which to work, thus perpetuating and solidifying the problematic historical literature on bilingual education. The central problem with this literature is that it lacks context. Kloss's work, for example, is often little more than a historical list of the significant dates and legislation regarding bilingualism in the United States. Without context there is no meaning, just a snapshot or detail; one only has the image of "a stuffed animal set on the floor of a bank," as William S. Burroughs once wrote about the lack of context in dreams.[20] To avoid this meaningless image, one needs to understand the political, social, economic, and cultural circumstances that brought forth the law, policy, or curriculum and to see how the larger social dynamics influenced education and foreign-language-speaking students. With all of the flaws in the historiography, it is time for a new blueprint of the polyglot boardinghouse to emerge, one that is national in focus so that the patterns of bilingualism can come to the fore and that the interactions among and between ethnic groups can be seen.

A New Blueprint

The notion of a "polyglot boardinghouse" promoted through bilingual education brings to the fore the debate over, as one English-only advocate put it, "what it means to be an American."[21] The definition of "American" and who qualifies for the title are historically contested issues. The United States—often thought of, at least by its citizens, as the preeminent liberal democracy—has continuously struggled with the granting of citizenship status to people who could possibly undermine the supposed unity of the nation, particularly people of the "wrong" gender, ethnicity, or race. A nation based upon liberal and democratic principles would suggest that the society would try to live up to those principles and, therefore, confer citizenship on those who were initially and wrongly denied that status. A liberal government, philosopher Barry L. Bull has rightfully noted, has the duty of facilitating the sort of "access" that is needed for "pursuing one's [vision of the] good." While there has been much progress in the granting of access, including the title of "American," to groups that have historically been marginalized, that progress has not been linear. As Rogers Smith has argued, U.S. naturalization policies cannot be characterized as a forward march toward inclusion; liberal and democratic ideals often mix with "ascriptive elements," creating an "overall pattern...of fluctuation" with regard to the meaning of "American."[22]

These "ascriptive" aspects that get thrown into the mix of deciding who qualifies for U.S. citizenship are often widely shared American attitudes that are outside of the Enlightenment principles of liberal theory, attitudes such as those concerning morality, nationalism, patriarchy, and white supremacy. Because schools frequently mirror the larger society, debates over the definition of "American" and the public attitudes and values that contribute to the debate are partially reflected in educational policies, especially those regarding bilingual practices. But, a wide lens is needed to see how these attitudes shaped policies. "If we stop history at a given point," E. P. Thompson has noted, "then... [there is] simply a multitude of individuals with a multitude of experiences. But if we watch these men over an adequate period of social change, we observe patterns in their relationships, their ideas, and their institutions."[23] These patterns are the key to historical inquiry, and examining them is a central aim of this book because, from a broad perspective, one can see certain patterns related to the debates over American identity and bilingualism.

This study argues that the patterns of bilingual education are closely related to the presence of public values in the United States. That is, how Americans define the ideal society and the good "American" alters the nature of schooling in general and bilingual education in particular, and frequently those values clashed with those of immigrant, indigenous, and colonized groups. American values, of course, evolved and changed between the 1840s and 1970s, which naturally brought forth new ideas about schooling and about the place of bilingual education in public schools. These shifts in public sentiments were sometimes the outgrowth of major events, such as depressions and wars, and sometimes related to larger social and intellectual trends, such as the rise of progressivism and scientism. In short, this book explores the ways in which cultural, social, economic, political, and educational trends influenced bilingual education in the United States.

The history of bilingual education in America's public schools is not simply a story about a back-and-forth struggle between ethnic groups and the English-speaking, Protestant majority, although, at times, that is at least part of the narrative. That story is too simplistic and ignores the multiple layers of power, imposition, and agency. There have long been subtle—at times, not so subtle—hierarchies of ethnic groups and languages in the United States, which underscores the need for some understanding of the interrelations between power and culture. In his work on the English working class, Thompson has demonstrated the intertwining of group culture and societal structures. A cultural identity, Thompson has argued, is not simply imposed on a particular group by the social forces of a dominant class, nor are marginalized groups fully autonomous. This study too explores the mixing of agency and structure in order to better understand the multifaceted levels of power.[24]

Like Thompson, some historians of education have examined the intertwining of structure, ethnic culture, and agency to capture immigrant experiences in schools. The "older" immigrant groups, for example, were often successful at implementing bilingual programs in the public schools, while the "newer" southern and eastern European immigrants rarely had public school programs to maintain their linguistic heritage. The nativist and racialist attitudes of Americans go a long way toward explaining these differences, but such structural inequalities and the imposition of English-only schooling do not take into account the choices, albeit limited choices, made by these less-favored peoples. The Poles, because of a devotion to their homeland's Catholicism, often sought cultural and linguistic maintenance in parochial schools, as did some Mexican Americans in the Southwest. In addition, many immigrant groups defined success and security in America as homeownership, a goal that required children to enter the workforce at a relatively young age. Therefore, formal schooling, even to maintain their linguistic heritage, was not always an essential endeavor for groups such as the southern Italians.[25]

Another issue of power embedded within this study relates to language itself. The philosopher Michel Foucault has noted that "in every society the production of discourse is at once controlled, selected, organised and redistributed according to a certain number of procedures, whose role is to avert its powers and its dangers." This control of language, Foucault has argued, is accomplished through "*exclusion*," which "relies on institutional support:...practices such as pedagogy—naturally—the book system, publishing, libraries,...[and] the learned societies."[26] These institutions were often places of language "exclusion" for foreign-language-speaking Americans. Obviously, the promotion of English in the public schools was a means of "controlling" the discourse of immigrants, Native Americans, and colonized Mexicans in the United States, but so too was bilingual instruction. Language is full of complexities, and foreign-language speakers in America actually spoke a multitude of dialects rather than a few unified, national languages. In fact, "dialects," as the linguist John McWhorter has persuasively argued, "are all there is"; typically, one "dialect"—often because of the powerful people who spoke it—was granted the status of "language" and, thus, became the "standard" or "official" tongue of the nation. The utilization of bilingualism in schools, therefore, meant that a particular variant of a language was given precedence over other dialects and that the leaders of bilingual programs essentially imposed—by defining what was to be the orthodox language—a foreign tongue on an ethnic community.[27]

Like language itself, bilingual education is quite complex. In a broad sense, it can encompass a number of very distinct types of programs and aims. To avoid the modern (yet anachronistic) view that only certain forms education are truly bilingual, dual-language instruction is defined broadly in this study. To capture a variety of foreign-language activities within American public schools, it is defined as the utilization of two languages to impart some form of knowledge: cultural, curricular, or linguistic. Bilingualism, as Haugen has noted, does not necessarily mean fluency in both languages; it merely requires that "the speaker of one language can produce *complete, meaningful utterances* in the other language."[28] The aim—or, in some cases, the unintended consequence—of America's elementary foreign-language programs was to make students bilingual, to have the ability to express themselves in more than one language.

It is hoped that this study, with its perspective of the past, is more than an antiquarian glimpse at an important social, cultural, and educational issue. The book is not intended to be an encyclopedic account of bilingual education; there are certainly many aspects of the issue that are not fully explored. Instead, the study maps out the landscape of bilingual schooling in America's past, its broad contours and elevations, as well as the uses of the terrain; it also invites other cartographers to fill in the details about the scenery. But, most importantly, it examines the events that changed the bilingual landscape, the subterranean shifts that brought forth mountains and canyons overtime. In his scholarship on moral education, B. Edward McClellan has noted that his work does not support a particular approach to character education, but rather seeks to broaden the discourse surrounding the issue. This book, I hope, will do the same for bilingual education. With the exception of using bilingual instruction for academic enrichment, I personally am not a strong advocate or opponent of any particular approach to bilingual education. Which method best serves the needs of English-language learners is a debate best left to parents, educators, policy analysts, and language scholars, not to historians. Although it will not answer those tough questions about how to educate the growing numbers of children in America's schools whose native language is not English, this study, I hope, will inform the issue—as McClellan's work does for moral education—"by offering perspective and suggesting a rough sense of limits and possibilities."[29]

Laying the Foundation for the Boardinghouse: The Context of Nineteenth-Century Schooling and Bilingualism

The dawning of the nineteenth century brought with it a new American president, one whose vision both foreshadowed and encompassed many of the changes that the United States was to undergo during the next several decades. Thomas Jefferson held out the promise of majoritarianism and egalitarianism to Americans. He stated in his 1801 inaugural address that "the will of the majority is in all cases to prevail," although "the minority possess their equal rights, which equal laws protect, and to violate would be oppression." While Jefferson's vision was complex and often tempered by the realities of political negotiations, it was clear that his notions of majoritarianism and egalitarianism depended upon homogeneity; only an agrarian society composed of intelligent, independent farmers, who spoke a common language, could create a nation of equals. Despite the fact that "faith in majority rule.... was constantly on his lips," America's third president thought he saw a tremendous threat to his democratic principles; what he saw was the rise of commerce in the new nation. Jefferson was leery of the budding manufacturing interests in the United States because he foresaw the "incompatibility of an immense proletariat and an equalitarian political democracy." The triumph of Jefferson's Republican Party in the national election of 1800 did not mean that "majority rule" and individualism would prevail over the drive for the elite control and centralization that the Federalists so desired. Instead, the friction between the two worldviews, as Joseph Ellis has pointed out, was "built into the fabric of our national identity."[1]

The manufacturing that Jefferson dreaded became an ever-increasing reality during the first half of the nineteenth century. The opportunities of burgeoning industries scented the air of the United States for the swarms of people all over the globe whose homelands were full of tainted nectar. Crop failures, floods, wars, persecution, and, especially, that fungus that blackened the potatoes of Ireland gave individuals from Asia to Europe the motivation to emigrate to America. And emigrate they did. Between 1831 and 1840 nearly 600,000 came to the United States; during the next decade over 1.7 million arrived. In the first six months of 1850 nearly 370,000 foreigners made the voyage to America. Although the Irish were the most numerous, many of the newcomers were from the non-English-speaking lands of Europe. The United States, by mid-century, was beginning to look like the "polyglot boarding house" that Theodore Roosevelt would later fear.[2] As more cultural, religious, and linguistic diversity washed upon U.S. shores, Jefferson's vision of a homogeneous society took on a new urgency. Building a uniform nation would require public education and, perhaps, a little of what Jefferson and his Republican followers most despised—those dreaded ideas of elite control and centralization—although it would be localism that would prevail for much of the century.

In 1779, Jefferson made a plea for public schooling to the legislature of his home state; four years later, he once again made a call for a system of education in his *Notes on the State of Virginia*. Jefferson noted that public education was a necessity in a republic, not only to create an intelligent citizenry for a nation that in a matter of decades would withdraw most of the restrictions on suffrage for white men, but also to prevent the country from becoming "perverted...into tyranny." Benjamin Rush, one of the signers of Jefferson's Declaration of Independence, also noted the need for a system of public schooling, but emphasized its homogenizing function. Rush, for instance, stated that public education was required in his state of Pennsylvania because "our citizens are composed of the natives of so many different kingdoms in Europe. Our schools of learning...will render the mass of the people more homogeneous and thereby fit them more easily for uniform and peaceful government."[3] The goal of public education, therefore, was to aid in the process of creating a unified society, a goal that was central to the Jeffersonian vision of America. The development a cohesive society through an educational system, however, was complicated by the non-English-speakers of the United States—originally those in the Northeast, Midwest, and the southern port cities—who hoped to use the newly created public schools to maintain their linguistic and cultural heritages. Many of them hoped, at least partially, to make America a polyglot boardinghouse.

"For the Protection of Society": The Common School and Nineteenth-Century America

Two decades after Jefferson left the nation's highest political office, a new president came to the fore to champion an egalitarian America even more forcefully than the third president could have ever contemplated. Andrew Jackson, the hero of the War of 1812 and former U.S. senator from Tennessee, noted in his last address as the seventh president of the United States that "[t]he planter, the farmer, the mechanic, and the laborer.... are the bone and sinew of the country." Yet, Jackson warned in 1837 that "they are in constant danger of losing their fair influence in the Government" because of "the power which the moneyed interest derives from a paper currency which they are able to control." Among many Jacksonians, gone was the fear of the urban proletariat whom Jefferson thought incompatible with a democracy of equals, but the distrust of the "moneyed" aristocracy persisted. As the "common" man's president, General Jackson had taken on the national bank and won over not only the laborers, but also Jefferson's yeoman farmers. By the Jacksonian period, suffrage had become nearly universal among white men, a pattern that had begun under the Jeffersonians. With the expansion of the franchise throughout the 1820s and 1830s, the "common" man was now voting, and he decided to cast his ballot for the "tobacco chewing" president and his political allies, which was worrisome to American conservatives. For one, Catharine Beecher, public school advocate and daughter of the Calvinist preacher Lyman Beecher, expressed her concern when she noted in 1835 that "these ignorant...adults are now voters, and have a share in the government of the nation."[4]

The tension between popular democracy and elite control in U.S. political culture continued in the Jacksonian Age, but instead of the Republicans battling the Federalists for influence over America, the Democrats wrestled with the Whigs in the second quarter of the nineteenth century. As in the earlier period, there was a great deal of ideological mixing. The Democrats promoted themselves as the party of the common folk, while the Whigs often catered to that "moneyed interest" that Jackson feared, which led to a potential dilemma for some American voters. "I am too smart to be a Democrat," stated the antagonist in a Henry Boernstein novel, but as "the child of poor parents, I was not born to be a Whig." While one would expect that a "common" party would be at the forefront of the movement to create "common" schools, this was not necessarily the case. Jefferson, that leader of small farmers, had made his plea for a system of schools, but public education did not become a reality under his influence or, for that matter, under the influence of his Republican Party. The environment was not yet ripe. Public schooling came into being after Jefferson, James Madison, and James Monroe exited the national stage, and the new manifestation of the egalitarian (and decentralizing) wing of Jefferson's party, the Jacksonian Democrats, failed to take the lead in establishing state-funded schools.[5] Instead, it was the conservative Whigs who brought "common" schools to the United States, but their notion of "common" was not a reference to the Jacksonian idea of ordinary folk. Instead, the "common" in common schools embodied the Jeffersonian ideal of homogeneity, a common culture and understanding to ward off the growing evils of the nineteenth century.

There was something very unsettling about the new century to those witnessing its many changes. For them, the eighteenth century had seemed much more cohesive and secure. From the mid-eighteenth century to the early decades of the nineteenth century, the tenor of American life was generally relaxed, at least compared to the tumultuous 1600s. New Englanders settled into their "comfortable" communities, and—although not living in close-knit villages like those in the North—Southerners experienced a "new stability," allowing Americans to have a level of "confidence in their society." The eighteenth-century village partially derived its stability through patronage and a web of intimate relations. Community members knew their position and deferred to their betters, creating a sense of cohesiveness. Iain Pears has captured this system of patronage in a novel about seventeenth-century England by noting that a man "was obliged to apply to others for support, as those beneath him must apply to him in turn. How else can any civil society continue to work, without a constant flow of gratitude from one to the other, high to low?" Social stability also emanated from the community's faith. Yet, religion had become more moderate by the eighteenth century. Even the rigidity of Puritanism had eased up a bit as the sect shed its isolationist position; Puritans eventually joined forces with other Protestant groups to create a united front against the growing threat of Catholicism. Although some of the strictness had been tempered, eighteenth-century Protestantism still provided communal stability and gave Americans a sense of confidence about their society.[6]

This newly found confidence about eighteenth and early-nineteenth-century America facilitated a leniency with regard to formal instruction, thus allowing for a variety of educational alternatives. Although the family remained the central educational agency, a hodgepodge of other institutions gave youngsters the rudiments of learning during the eighteenth century and continued to do so during the formative decades of the new republic. In the northern states, town and district schools, often partially supported by the public coffers in Massachusetts, sat alongside dame schools, charity schools, and a host of other "private-sector" educational institutions in the early national period. "Private" and "public" schooling, however, were not entirely distinguishable until the early decades of the nineteenth century. Many educational institutions were funded by a mix of public largess, private donations, and tuition fees. During the eighteenth and early-nineteenth centuries, this mixed system of educational institutions served the needs of the American population for the most part, particularly those of the growing merchant classes. The mixed nature of institutions for formal learning, however, led many school reformers to conclude that education was on the decline in early-nineteenth-century America.[7]

This anxiety over educational regression was coupled with a growing uneasiness about changing economic, demographic, and cultural patterns, an uneasiness that precipitated the development of common schooling in the United States. Largely gone were the days of the stable, hierarchical village in which deference was the norm; mobility and universal white suffrage had chipped away at such traditions. But, more important, the Industrial Revolution made its way to the Northeast by the 1830s. Textile mills and other industries dotted the landscape in states such as Massachusetts and New York. By 1850, the United States had more steam power than any of the countries of Europe, including Great Britain. American agriculture too was going through a transformation. Crop specialization, as opposed to subsistence farming, was becoming more commonplace, and Jefferson's yeoman farmers were gradually becoming a dying breed. A revolution in transportation allowed the agricultural sector to send raw materials to the mills and to feed the growing populations in the industrial towns. During the first half of the nineteenth century, therefore, America was changing from a traditional, mercantilist society to one that was wedded to laissez-faire capitalism.[8]

The industrial boom in the Northeast during the second third of the nineteenth century created demographic shifts. The emerging mills facilitated a general march to the industrial centers. Rural folk increasingly began to move to the manufacturing towns that offered economic opportunities, thus ushering in American urbanization. Massachusetts, for example, had three towns with a population over 10,000 in 1810; by 1840, it had nine, and two decades later the Bay State had twenty-six urban areas. The city was both a source of hope and anxiety for many nineteenth-century Americans. The urban center was a symbol of progress and civilization, but if its ills were left unchecked, it could undermine the stability of the nation. The central goal was, as historian Thomas Bender has argued, to create a "harmony between city and country" in order to keep the potential dangers from growing out of control. America's industrial and urban development was also a draw for foreigners. The Irish, for example, came to the United States as a means of survival, particularly after the potato famine had destroyed their traditional way of life; about a million people had died by mid-century as a result of the crop failures in Ireland. The immigrants arrived in large numbers during America's industrial explosion, with nearly five million new arrivals in the second third of the nineteenth century. The vast majority of the newcomers—ranging between 75 and 97 percent—was from northern and western Europe, and in the United States many found work in the manufacturing sector. Irish immigrants gradually replaced the native-born in the textile mills of the Northeast because they—unfamiliar with the traditionally slower speed of spinning work—were willing to work at the faster pace that the new industrial order demanded.[9]

While the Americans of the eighteenth century found cohesion in community life and faith, the industrialism of the nineteenth century eroded much of that stability. The rise of capitalism brought with it "absentee ownership." This very impersonal form of economic relations was vastly different from the almost familial types of patronage and commerce that marked much of the eighteenth century. Because the new industrial age was characterized by impersonal economics, American capitalism was moving away from "a personal moral code," which allowed for very immoral actions among the emerging industries. But, nineteenth-century America would not be without any sort of ethical guide. Increasingly, government agencies stepped in to ensure that people acted as they should. Public education, a new, albeit local, governmental function, was in the forefront of the movement to restore America's cohesion, stability, and morality. That great common school champion, Horace Mann, stated in 1848 that the public school will "wield its mighty energies *for the protection of society* against the giant vices which now invade and torment it...[;] there will not be a height to which these enemies of the race can escape, which it will not scale, nor a Titan among them all, whom it will not slay."[10] School leaders, so it seemed, had a great deal of faith in the power of their newly created institutions of learning.

It was within this context of rapid social change that the movement for common schooling found a population that was eager to hear what school reformers had to say. What they said was largely what all portions of the audience wanted to hear; the advocates of common schooling and public high schools promised all things to all people. The common school reformers, such as Mann in Massachusetts, Henry Barnard in Connecticut, Calvin Stowe in Ohio, and Caleb Mills in Indiana were a relatively homogeneous lot; they often had formal or familial ties to leading Protestant churches, were politically conservative, and came from the middle classes of American society. From their newly created state-level positions as secretaries of education, many common school advocates argued that education was a panacea. Public schooling would simultaneously promote industrial growth and solve the problems associated with industrialization and urbanization, such as crime and pauperism. This "implausible ideology" suggested that industry would be bolstered by public schools through the training of the younger generation in the habits demanded by industry, while social vices would be greatly curbed by inculcating Protestant morality and self restraint.[11]

For the school reformers, overcoming the social ills associated with industrialization and urbanization required a renewed social cohesion, a cohesion that was supposedly lost in the early decades of the nineteenth century. The common schools, however, would not replace the communal morality of the late colonial era, the morality that emanated from the shared personal piety of individual members; much of that was lost in the new century. Yet, "as the rules of private belief loosened," one scholar has stated, "the rules of public behavior tightened," and nineteenth-century America required a strict conformity to social rules. This conformity in American life was duly noted by Alexis de Tocqueville during his famous trip to the United States in the early 1830s. Tocqueville noticed that "public and private life is constantly muddled together" in a democracy and was concerned that "there is no freedom of thought in America." From the Tocquevillian perspective, Americans seemed to have bought into Marcus Aurelius' aphorism: "What is no good for the hive is no good for the bee."[12]

"Public behavior," historian Robert Wiebe has insightfully noted, "once merely signifying the state of the soul, had become the substance of virtue," and inculcating that public virtue became the central function of the common school.[13] Of course, the newly created public schools gave students the rudiments of reading, writing, and arithmetic, but moral training reigned supreme in the schoolhouse. As many nineteenth-century Americans came to believe that "only absolute rules rigidly adhered to could provide a reliable guide to behavior and protect against the enormous temptations of the day," schools emphasized those rules, rules that largely paralleled pan-Protestant behavior. "The school," stated the trustees of New York City's nineteenth ward, "is no place for a man without principle," and the textbooks and teachers of the common school era reflected this focus on public virtue, stressing honesty, hard work, thrift, temperance, and obedience. In their texts, children learned that God was always watching, a subtle lesson to ensure "self-control." But the God that students learned about was a Protestant God. Protestant beliefs and Protestant scriptures were what students encountered in the "common" schools; Catholicism was "depicted not only as a false religion, but as a positive danger to the state."[14] The "American people," one superintendent in Pennsylvania stated, "are a religious people," and, therefore, the Protestant Bible and prayers were a necessity in the common schools; the scriptures, however, were read "without note or comment" so as not to antagonize any particular Protestant sect. (Perhaps poking fun at the common school practice, Mark Twain once noted that "religious tracts" were thrown off Mississippi River steamboats and picked up *without comment.*")[15]

Public schooling began in the Northeast, and, although some children had been attending school in America for centuries, the common school movement provided an impetus for formal education to reach more youngsters. In Massachusetts, for instance, nearly 1,500 new public schools were created in the two decades between 1840 and 1860. In urban areas like New York, a scattered collection of independent charity schools laid the groundwork for consolidating a public school system throughout the city. The idea of common schooling spread to the new states and territories as the New England population moved westward, but the South lagged behind until after the Civil War. Common schooling did not take root in the South because of demographic patterns—especially the low population density among whites—and the elites' resistance to interference with their traditional way of life. As a result, only a handful of southern port cities developed public school systems prior to the 1860s. During his visit, Tocqueville noted that in the North, where the drive for public schooling was beginning, the citizens were more moral and made "better choices" at the ballot box, but in the South, "where education is not so widespread and where the principles of morality, religion and freedom are less happily combined, one observes that the aptitudes and virtues of government leaders are increasingly rare."[16] Like Mann during the next decade, Tocqueville had a great faith in the molding power of education.

Yet, the common school movement was more than just a benevolent desire to bring learning to the masses, it also contained an ideology. Key to the ideology was social cohesion, a cohesion that emanated from what Catharine Beecher called "moral and religious restraints." But developing an educational system that promoted social stability and uniformity demanded that schooling be a shared experience—"common" to all segments of the society—which necessitated a degree of centralization and standardization in the realm of education, goals that were key to the Whig vision of America. The reformers' desire for the centralization of schooling and the standardization of curricula and teacher training, however, encountered sharp resistance at times. That is, the designs of an elite group of reformers occasionally clashed with popular sentiment. Many recognized that the centralization embedded within the common school ideology threatened local control, which generated resistance to the newly created institutions. In addition, the working classes, particularly skilled craftsmen, sometimes rejected the social control and industrialization aims of school reformers, while the notion of a common school that promoted Protestant morality met with resistance from the growing Catholic population. As with all movements for educational change, common school reforms had to be negotiated, and, therefore, Mann and other reformers attained only a partial victory. The new concept of publicly funded common schools was laid on top of the old structure of district schools, a structure that resisted centralization.[17]

Although the common school advocates pushed for centralization, local control, which took root in colonial America, reigned supreme for much of the nineteenth century. Jefferson and Jackson's resistance to elite control could not easily be quelled. "America," one historian has stated, "was 'born free' of an institutional framework," such as a monarchy, but "'[b]orn free' in America meant born in pieces.

So little resisted the pulls of local attachment." Localism shaped the newly created public schools; the curriculum, the teachers, the schedule—all were determined by the needs and desires of the local population. Even in the urban areas where centralization was more prevalent, the citywide system allowed each district a certain amount of autonomy, and the locals of each ward tailored education to their own needs. The opening up of the western lands also encouraged localism. Settlers, if dissatisfied with their communities back East, could start a new one out West. It was partially this pattern of localism in nineteenth-century America that allowed bilingual education to flourish in what historian John Rury has called these "profoundly local institutions."[18]

Polyglot America

Linguistic diversity has always been a feature of the territory that would become the United States. Prior to the Revolutionary War, German, Dutch, French, Spanish, English, and—most importantly— a host of indigenous languages could be heard throughout the lands of America. Once the colonies broke from mother England, polyglottism continued; no national language came into being. While there was a great deal of talk about giving English official status, the leaders of the new nation often associated such policies with monarchical rule, the very thing the states had just fought a bloody war to end. Although there would be no official language and while many national leaders actually encouraged multilingualism (for practical purposes), English, the dominant language on the eastern seaboard, was largely taken for granted as the language for public affairs.[19]

The laissez-faire attitude toward multilingualism in the new nation was partially an outgrowth of the fact that, as linguist Shirley Brice Heath has noted, "language was not yet a national ideological symbol." The United States did not couple language and nationalism as strongly as did the European states that were seeking nationhood. For those who were trying to throw off the yoke of some great empire, historic mother tongues became not only the means of communicating the ideals of independence, but also the means of identifying the members of potential nation-states. While language and nationalism were not as closely bound in America, the promotion of *American* English was a means of distinguishing the new nation from its former parent country. Noah Webster, that former New England schoolteacher whose name is forever linked with the American language, noted that "men whose business is wholly domestic have little or no use for any language but their own"; their "own" language in the former colonies was largely a particular variant of England's tongue. As Webster compiled his spellers and dictionaries in the late-eighteenth and early-nineteenth centuries—those books that schooled generations of children linguistically—American English became less flowery than the King's English and incorporated many indigenous words and phrases, which those from Great Britain considered "barbarisms."[20]

While the comparatively small amount of linguistic nationalism allowed for multilingualism to flourish in the United States, so too did the vastness of the nation. Jefferson's purchase of the Louisiana Territory brought huge tracts of land under the control of the new nation, land west of the Mississippi River that would eventually become Iowa, Missouri, Arkansas, Nebraska, and South Dakota, as well as parts of eight other states that stretched from Louisiana to Montana. After the war with Mexico in the 1840s, the United States took what is now the American Southwest. By the 1850s, therefore, the United States was vast and quite "open." Yet, "[o]pen land," as Wiebe has noted, "accentuated a tendency to preserve particular beliefs, to prize homogeneity, to leave all antagonists holding some position triumphant. Differences were spread across space rather than managed within it."[21] Hardly anything in this vast land, a land nurtured on localism, could prevent language minorities from maintaining their cultural traditions.

Nothing, of course, contributed to America's linguistic diversity more than the large numbers of people in the nation who spoke a language other than English. Between 1850 and 1880, millions of foreign-language speakers emigrated to the United States (see table 1). Numerically, those from the Germanic kingdoms, states, provinces, grand duchies, principalities, and free cities dominated. By 1880, the United States could count nearly 2 million foreign-born Germans among its inhabitants. There were many linguistic minorities who were not immigrants, however. The foreigners' American-born children, while not technically immigrants, spoke languages other than English. So too did the hundreds of thousands of American Indians. In 1870, there were nearly 384,000 Native Americans living in the United States and its territories, the majority of whom resided west of the Mississippi because of the removal policies of the federal government during the first half of the nineteenth century. Of the hundreds of thousands of Native Americans, only about 30,000 remained in the states east of the Mississippi in 1870.[22]

Table 1   Foreign-Language-Speaking Immigrants in the United States by Decade[i]

Origin	1850	1860	1870	1880
Germany	584,000	1,301,000	1,690,000	1,967,000
France	54,000	110,000	116,000	107,000
Norway	13,000	44,000	114,000	182,000
Sweden	4,000	19,000	97,000	194,000
China	760	35,000	63,000	104,000
Mexico	13,000	27,000	41,000	68,000

Note:
i. Census information can be somewhat contradictory, making it difficult to interpret. For example, the census for 1860 lists the number of Chinese in America as 35,565, while the 1870 census states that there were 34,933 Chinese in the United States in 1860. In addition, after 1870 the census reports break foreign immigration into racial categories, such as "white," "colored," and "Indian." To avoid much of this confusion, the number of immigrants for each decade is rounded off. The "Germany" category includes German-speaking areas—such as Prussia—that would become part of Germany proper after 1871; this category does not include other German-speaking nations such as Austria and Switzerland.

Source: J. D. DeBow, *The Seventh Census of the United States: 1850* (Washington, DC: Robert Armstrong, 1853), xxxvi; Joseph C. G. Kennedy, *Population of the United States in 1860; Compiled from the Original Returns of the Eighth Census* (Washington, DC: Government Printing Office, 1864), xxviii; Francis A. Walker, *The Statistics of the Population of the United States, Embracing the Tables of Race, Nationality, Sex, Selected Ages, and Occupations*, ninth census, vol. 1 (Washington, DC: Government Printing Office, 1872), xvii, 336–342; Department of the Interior, Census Office, *Statistics of the Population of the United States at the Tenth Census (June 1, 1880)* (Washington, DC: Government Printing Office, 1883), 492–495; Heinz Kloss, *The American Bilingual Tradition* (1977; reprint, with a new introduction by Reynaldo F. Macias and Terrence G. Wiley, McHenry, IL: Center for Applied Linguistics and Delta Systems Co., 1998), 11.

Immigrants and their children constituted a majority of the foreign-language speakers in America, and it was this "older" wave of immigration—particularly the Germans and Norwegians—that was most active in establishing bilingual public schools in the Northeast and Midwest before the 1880s. While the South was not the primary destination for European immigrants, some did make their home in a handful of southern port cities and, in some instances, helped set up bilingual schools in the region. Prior to the Progressive era, the Germans were the most numerous group of immigrants speaking a foreign tongue in the United States. Their emigration was quite uneven, however; they came from particular regions at particular times. In the first half of the nineteenth century, German immigrants often emigrated from the western lands, especially the southern grand duchies and kingdoms such as Baden and Württemberg. Increasingly, immigrants from the northwest portion of the Germanic lands also arrived in large numbers. After the 1860s, however, Germans from the northeastern areas—particularly from Prussia—began to emi-grate to the United States. In general, the Germans who arrived in the United States during the second third of the nineteenth century tended to come from the middling segments of society and frequently came in family units. By the latter part of the century more Germans were arriving as individuals from the eastern territories and were fre-quently from the lower classes. Like the Germans, the Norwegians too initially came in family groups, and by the 1890s, more solitary Norwegians made their way to America.[23]

What this lopsided pattern of emigration suggests is that immi-grants left their native lands for specific reasons, reasons that were intimately linked to regional contexts, social-class conditions, religious beliefs, and the political climate. The United States, although attrac-tive to emigrants because of its industrial development and open land, was not a magnet simply pulling the poor peasants of Europe to new shores. As circumstances in Europe changed, America became a viable option for some, but not for others. Early Norwegian emigrants came to America in search of property after population growth and chang-ing economic patterns in Sweden—Norway gained its independence in 1905—made land ownership a difficult endeavor. The Norwegians who arrived after the 1880s often came from urban centers in the old country and frequently sought money in America, not land like their older brethren. German immigrants also came for a variety of rea-sons. In the 1830s, as economic conditions changed in the southwest-ern regions, many middle-class Germans, sometimes referred to as the "Thirtiers," came to the United States as a means of retaining their wealth and status. Several groups of "Old Lutherans" left the north-eastern sections in the second third of the nineteenth century after the 1817 state-supported merger of the Lutheran and Reformed Churches threatened their religious traditions and autonomy. The political refu-gees arriving in the late-1840s, the so-called "Forty-eighters," had left Europe after the failure of their revolutions of liberation; they right-fully feared a repressive backlash. In addition, many of those coming from Hesse-Darmstadt in the 1850s were escaping the growing unem-ployment of that region. Failing crops and rising prices also facilitated the exodus from western Germany in the 1850s. Johann Bauer, who emigrated from Baden in 1854, hoped to improve his lot in life by com-ing to America, and he was not disappointed. Writing to his family in the fatherland, Bauer noted in 1855 that "you can see how thriving the conditions are here, & that with a bit of hard work & stamina it is easier to get ahead than in Germany."[24]

While the decision to emigrate was based on a variety of circumstances, coming to America, for whatever reason, was no small undertaking. Foreign-language-speaking immigrants set sail from northwestern European ports—especially Hamburg for the Germans—and made their way to cities such as New York and Baltimore. Some Germans headed for the Midwest traveled directly to New Orleans in order to catch a steamer up the Mississippi; like the river itself, Herman Melville wrote in the mid-nineteenth century, the immigrants traveled "helter-skelter, in one cosmopolitan and confident tide." The voyage from Europe to the United States was long and arduous. One traveler in the 1850s recommended that emigrants spend extra money for a cabin on the ship instead of remaining in steerage because "[i]t is no small matter to live in a dark room with ca. 160 to 200 people for 40 to 50 days." Additional funds, of course, were hard to come by for many. German immigrant Wilhelmina Stille, for instance, noted in the 1830s that "the trip cost so much that I didn't believe it myself, the Americans aren't ashamed to overcharge the Germans."[25] Greater access to steamships made the journey shorter for the immigrants who arrived later in the century, but traveling was still hard. In his tale of American immigrants, novelist Howard Fast has captured the torture of crossing the Atlantic in the 1880s: "Misery absorbed them. Nausea absorbed them. The agony of their stomachs absorbed them." There were other elements of the voyage that absorbed the immigrants as well. Nikolaus Heck wrote in 1854 that "there were 300 of us when we went on board the ship. But when we left, our numbers had increased to the millions. And these countrymen were called lice." Illness too absorbed them, as it was a perennial danger. Cholera, for instance, was present on over 30 percent of the ships from Hamburg to New York in 1853. Fifteen years later, the New York State Commissions of Emigration inspected the *Leibnitz*, which had traveled from Hamburg to New York, and discovered that of the "544 German passengers, 105 died on the voyage, and three in port."[26]

Although coming to America was a challenge, the ultimate destination of the newcomers was sometimes a result of why they left and from where they emigrated. Matthias Dorgathen, a immigrant from Germany's Ruhr District, left his homeland when coal mining became less profitable after the depressions of the 1870s. Not surprisingly, Dogathen settled in the coal districts of eastern Ohio. Similarly, the Norwegians who left Europe in search of land put down roots in America's upper Midwest, an area that had abundant land and that, climatically, was not unfamiliar to those from northern Europe, including its harsh winters. In fact, the older immigrant groups tended to settle in locations that were somewhat similar to the environments of their homelands. Those who chose locations that were climatically different from what they were used to sometimes found it difficult to adjust. After emigrating from Germany, Carl Berthold spent several years in the South. In Louisville, Kentucky, Berthold found that "the heat is too oppressive," and in Jackson, Mississippi, he noted that "the heat is much worse than in the North." Like the land-hungry Norwegians, the German Old Lutherans, after being persecuted in the fatherland, sought out sparely populated parts of the United States in order to ensure their religious freedom and to build a community of like-minded believers.[27]

Why the emigrants left their homelands also partially determined their social and cultural activities in the United States, which was particularly true of the German Forty-eighters. Those who came to America after the European revolutions in 1848 were a secular and liberal lot. Having disavowed organized religion, these "freethinkers" were often members of the gymnastic societies, the *Turnvereine*, that helped organize revolutionary activities in the mid-nineteenth century. The Turner movement emerged at the beginning of the nineteenth century in Berlin under the charismatic leadership of educator Friedrich Ludwig Jahn. While "Father Jahn's" focus was on physical education, the Turner societies quickly became political associations. In the German provinces, the aim of the revolutionary Turners and other radicals was to bring forth a unified Germanic nation, one that embraced freedom and, therefore, threw off the yoke of the princely tyrants. After the uprisings failed to achieve a lasting liberal *Deutschland*, many Forty-eighters fled to the United States. America was the logical destination of the highly educated, liberal revolutionaries; it seemed to be an idyllic democratic society that was committed to freedom and tolerance. But, as many immigrants quickly discovered, the United States was no land of milk and honey. After the Forty-eighters encountered America's nativism, slavery, and restrictive blue laws, they sometimes became disillusioned with the United States. The liberal German Americans, therefore, sometimes became active in politics to challenge conditions they deemed unjust. Some even founded the town of New Ulm, Minnesota, in order to live up to the Forty-eighter ideals. The liberal Germans also were disenchanted by the lack of education among earlier waves of German immigrants. That "the German element...stood at a very low educational level," inspired many "educated men" from the 1848 revolutions, such as Boernstein in St. Louis, to "elevate German Americans" by establishing reading rooms, theaters, newspapers, and other cultural endeavors.[28]

This disillusionment with America was not confined to the liberal revolutionaries from Germany. All sorts of immigrants found that the United States was not the utopia that some imagined it would be. Norwegian immigrant Andreas Hjerpeland, for instance, wrote in 1873 that "America is no paradise, but it cannot be denied that it is a good land." Similarly, the Old Lutheran Pritzlaff commented in the 1840s that "America is a good country, it blooms under God's blessing, but it bears thorns and thistles as well." The United States, as it turned out, was not always a land of dreams; for some new arrivals, it was a land of nightmares. Boernstein, for instance, made note of "the earnest and sad feelings which seize on all immigrants in view of their new home." What brought on the sad feelings was often the dominance of English in the United States, which left the foreign-language speakers "silent, bereft of voice and will."[29] The English language was utterly frightening to many immigrants, which caused a great deal of anxiety among those who did not know the tongue. O. E. Rolvaag's *Giants in the Earth*, a novel about Norwegian immigrants, captured this fear when a central character, upon arrival in the New World, was "confused and bewildered by the jargon of unintelligible sounds," sounds "that did not seem like the speech of people." A Norwegian arriving in the early-twentieth century admitted she was "a bit nervous...even though I can manage the language. It is much worse for those who can't talk."[30] Numerous other immigrants expressed anxiety over the language of the United States. One German immigrant in the 1850s noted, "You can easily imagine how things are when you first come to a foreign country and can't speak the language, how things are *at first* and how one feels." How they often felt "at first" was what Heinrich Möller expressed in the 1860s: "I don't like being in America very much, since I don't like the English language." A few years later, however, Möller had become an Anglophile, admitting "I like to speak English better than German."[31]

Partially as a means of overcoming this anxiety about the English language, immigrants often settled in ethnic enclaves, in both cities and rural areas. "In America," a student of immigrants observed, "it is language and tradition...that unites the foreign populations. People who speak the same language find it convenient to live together." Immigrants themselves often expressed similar notions. Per Hansa, the protagonist in *Giants in the Earth*, stated that he and his fellow settlers "want only *Norwegians*" in their part of the Dakota Territory, noting that "we sift the people as they pass through here—keep our own, and let the others go." One German immigrant in New York felt some sense of security upon arrival in the 1850s because she was able to find a "German boardinghouse" and, later, an ethnic neighborhood where she was "among nothing but good people who are all Catholic and German."[32] Ethnic enclaves were a valuable source of support for the newly arrived; they could communicate in their native tongues and live among others who shared their anxieties about their new homeland. Not all immigrants, however, settled alongside of their fellow countrymen, which, naturally, sped up the Americanization process. Möller, for instance, eventually put down roots in an area of Pennsylvania that was non-German. By 1886, he noted that "my children...can't write German very well, and mine is almost as bad, too, since I sometimes go for a whole year without writing any German." As the Czech novelist Milan Kundera—himself an émigré living in France—has pointed out, "people who do not spend time with their compatriots...are inevitably stricken with amnesia."[33] To resist this amnesia and to calm their anxiety over language, many nineteenth-century immigrants attempted to maintain their cultural and linguistic heritages in a variety of ways, including through the establishment of numerous foreign-language institutions.

America's Polyglot Institutions

When the American tendency toward local rule was coupled with the foreign-language speakers' close proximity to one another in ethnic neighborhoods and settlements, hardly anything prevented cultural and linguistic maintenance among the immigrants in the North, South, and Midwest prior to the 1880s. Schools, of course, were not the only vehicles for the maintaining of cultural and linguistic traditions among the immigrants. The foreign-language press, a host of societies, and immigrant churches also helped foreign-language speakers to hold on to their mother tongues in a predominantly English-speaking nation.

The immigrant press was fairly substantial prior to the Progressive era. From Maine to Louisiana—and nearly everywhere in between—foreign-language newspapers were established. The foci of the papers, of course, varied; some were religious in nature, while others tended to be political, but many were vehicles for education and linguistic maintenance. The Acadian press in the Northeast fought against the Americanization of its French-Canadian readers. Similarly, the Louisianan Creoles—as well as many immigrant Acadians—had papers such as the *Abeille*, which was established in 1827, to preserve the French language. Norwegians, too, had a flourishing press. Papers such as *Budstikken, Decorah-Posten, Faedrelandet og Emigranten, Folkebladet, Norden, Skandinaven,* and *Verdens Gang* were all founded in the upper Midwest prior to 1880. As developed as the French and Norwegian papers were, it was German that came to dominate the foreign-language press in the United States. The first German-language newspaper in America, the *Philadelphische Zeitung,* was published in 1732. Ironically, Benjamin Franklin, that critic of the German language in Pennsylvania, helped with the establishment of the paper. But, it was during the nineteenth century that the German newspapers became ubiquitous. By 1860, the German-language press could boast about around two hundred papers in operation, several of which were dailies, particularly in cities such as New York, Philadelphia, Baltimore, Chicago, Cincinnati, Milwaukee, and St. Louis. The numerical strength of the German immigrants, obviously, was what supported this vast foreign-language press, but so too did the presence of the highly educated Forty-eighters, many of whom were affiliated with German papers. The *Baltimore Wecker,* for instance, was edited by well-known Forty-eighters such as Karl Heinrich Schauffer, August Becker, Karl Gottfried Beck, Wilhelm Rapp, and Franz Siegel. In 1850, Boernstein took over the editorship of St. Louis's *Anzeiger des Westens,* a daily, as historian Carl Wittke has noted, that was considered "the leading German journal in the country." Boernstein noted that the educated immigrants were "new additions to the German element [who] worked like a ferment, bringing intellectual strength and clarification after quiet brewing." The cultural endeavors of these fermenting Forty-eighters, including their newspapers, were devoted to keeping "German pure and to cleanse it of the Anglicisms that had contaminated it."[34]

In addition to his editing of the *Anzeiger,* Boernstein was intimately involved with a variety of German clubs and societies, the aim of which was to advance the education, language, and culture of German Americans. For example, he helped found reading clubs, the German Women's Association, and the Philodramatic Society. Boernstein also lent his leadership skills to the schools of St. Louis's Society of Free Men. By far, the most well-known of the German associations were the gymnastic societies (*Turnvereine*) and the men's singing clubs

(*Männerchöre*), both of which were common throughout the nation. As imports of the revolutionary Forty-eighters, the *Turnvereine* were more than mere gymnastic clubs; they had a political outlook, one that stood in opposition to American injustices, such as slavery. By the 1850s, several Turner societies were part of a national association, the *Sozialistischer Turnerbund*; the organization saw itself as devoted to education and justice. The *Männerchöre* may not have had the political aims of the Turner societies, but with their devotion to the Fatherland's music and language, their educational goals were just as profound. The men's choirs were also ubiquitous. Writing from Connecticut in 1855, German immigrant Martin Weitz was excited to tell his family that "[e]ven we in Rockville...have 2 singing clubs." "With our German singing," Weitz noted, "we earn great respect"; Rockville also began a *Turnverein* in 1857. In some ways, the immigrants, particularly the numerous German Americans, were paralleling a larger nineteenth-century American trend of forming associations, a trend Alexis de Tocqueville discovered during his famous tour of the United States. "Americans," Tocqueville noted in the 1830s, "constantly unite together." They have "associations to which all belong but also a thousand other kinds, religious, moral, serious, futile, very general and very specialized, large and small. Americans group together to...found seminaries, build inns, construct churches, distribute books....[and] establish hospitals, prisons, [and] schools." Perhaps wryly, Tocqueville also pointed out that if Americans "wish to highlight a truth or develop an opinion by the encouragement of a great example, they form an association."[35]

Like the ethnic associations, the foreign-language churches sought to maintain the traditions of the immigrants. In immigrant parishes, the Catholic Church was well-known for its preservation of foreign tongues in the United States. French Americans, whether Creole or Acadian, tended to be Catholic, as were many Germans, especially those from the southern provinces. By the early-twentieth century, over fifteen-hundred Catholic parishes used at least some German in their homilies. The Methodist and Lutheran Churches in the United States were also actively involved in maintaining the languages of immigrants. Numerous Norwegian Lutheran congregations, affiliated with a few different synods, were founded in Illinois and Wisconsin during the 1840s. German Americans also belonged to Lutheran congregations, and Cincinnati's North German Lutheran Church, founded in 1838, vigorously sought to preserve the German language. Several of the German Lutheran churches in the United States were part of the large and powerful Missouri Synod, and quite a few of the synod's congregations actively sought to protect the German language from America's English influence.[36]

This diversity among institutions for linguistic preservation brings to the fore two critical and related issues regarding the foreign-language speakers in the United States. The first is that America's immigrant communities were quite fragmented, regardless of whether or not the members spoke the same language. American immigrants, for instance, divided themselves over a host of religious issues. The German Americans were far from monolithic when it came to matters of faith. The division between the *Kirchendeutsche* and the *Vereinsdeutsche*—colloquial terms for "church Germans" and secular "club Germans" at the turn of the twentieth century—was obviously vast. At times, these two types of Germans had very little tolerance for one another. In 1858, Lenz decried the lack of religion among his fellow immigrants by noting that "there's too much godlessness and most of all among the Germans who...scoff at God and His justice." Another "church German" expressed similar feeling decades later, stating that "over half the English-speakers don't go to church.... never even heard of religion. And many Germans are the same. You can tell from the new immigrants that they don't care much about the church, since they don't join congregations like they used to."[37] By contrast, secular German Americans often found their religious countrymen "absurd." Clemens Vonnegut, a leading freethinker in Indianapolis, noted that, unlike the superstition of Christians, his "belief is built on reason, observation of nature, history and experience." In the 1850s, the German freethinkers in Louisville, Kentucky, and Richmond, Virginia, called for the abolition of all religious activities in political life, and demanded that religious property, including that of their fellow Germans, be taxed. Because of the oppression in Europe, many of the secular Germans in America were virulently anti-Catholic. Boernstein's writings depicted Catholic priests, particularly Jesuits, as corrupt and sinister. Another St. Louis German, socialist Ludwig Dilger, shared Boernstein's perception. "This country," noted Dilger, "is ruled by priests and petticoats. The whole pack of Jesuits that Bismarck chased out of Germany have made themselves at home here and are living high on hog, rolling in clover."[38]

Of course, immigrant groups were not only separated into believers and non-believers; the ethnic communities were also fractured into a multitude of faiths and sects, and serious divisions existed among and between the religious groups. Although many Norwegians were

Lutherans, this did not prevent the Norwegian Americans from splitting into a variety of congregations and synods, some of which were quite unorthodox by opposing the state-sponsored Lutheranism of their homeland. The German Lutherans also established churches to suit the needs of their members, thus creating a fair amount of diversity among congregations. In addition to the variety within religious denominations, immigrant groups sometimes represented a number of religious faiths. The German Americans, for instance, were not only Protestants, but also Jews and Catholics, and these distinctions, at times, brought forth conflicts within ethnic groups. During the nineteenth century, tensions between Catholics and Protestants ran high, especially over issues of moral education in the public schools. Despite their shared fatherland—at least after 1871 when the German lands united into a single nation—Protestant and Catholic German Americans could not reconcile their divergent beliefs. Catholics saw Protestants as "heretic[s]" because for them "there is but one church...and...it is the Catholic Church." Protestants and secularists, by contrast, depicted Catholics as an undemocratic "papist gang" who could not fully accept the tenets of the United States' political system.[39]

Religious beliefs, however, intersected with Old-World regionalism, which, in turn, also was a partial reflection of social class; both elements—regionalism and social class—had a divisive impact on ethnic groups. Germans' regional loyalty, in addition to their spiritual views, proved powerful. A German from northern Hanover noted that in his American town "our Germans are East Frisians, pretty terrible people when it comes to religion not only in this country but also in Germany." Immigrants frequently thought in Old-World regional terms. Even after Germany's unification, one German in St. Louis noted to his family that the city was full of "High Germans, Bavarians, Swabians, Badenese, Hessians, Saxons, Swiss, [and] Hungarian Germans," but no Westphalians like himself.[40] Rolvaag made note of tensions between Norwegians from Helgeland and those from Trondhjem, although Per Hansa, the central character in his novel, was willing to set aside the Old-World rivalry with the Trönders in order to survive in America. Like religion, regionalism overlapped with social class, which also divided ethnic communities, particularly when it came to issues such as labor unrest and socialism. Johann Witten, for instance, had no tolerance for "socialists" who "stir up the people," while his fellow countryman, Dilger, was so devoted to the movement that he predicted that by the turn of the

twentieth century "we'll have a Socialist president and *Congress*."
A handful of German liberals in the 1850s publicly demanded the
"[a]melioration of the condition of the working class...by lessening
the time to work to eight hours," a demand that did not sit well with
many Germans or native-born Americans. Moreover, Boernstein—
although somewhat sympathetic to the plight of the poor—clearly felt
superior to lower-class Germans who "stood on the same intellectual
level as the Irish...; instead of doing homage to whiskey as the Irish
did," these German peasants "regarded the beer mug and the tobacco
pipe as the highest goods of this earth." [41]

In addition to within-group divisions, the diversity that existed
among institutions for language preservation brings to the fore
a second, yet similar, issue, namely that ethnic groups in America
were often invented entities. In his study of nationalism, Benedict
Anderson has persuasively argued that nation-states are nothing more
than "imagined communities," communities that have been deliberately concocted and cultivated by leaders of nationalist movements.
Historically, language has been a central tool of nationalists, who
often promote "national print-languages" to bind citizens to the state
and to each other. [42] In the United States, immigrant groups were, in
a sense, also "imagined communities." That is, despite the distinctions regarding religion, regional affiliation, and social class, broad
groupings—such as "German" or "Scandinavian"—utterly simplified
immigrants' own loyalties and promoted largely made-up categories
of peoples.

For many, "Norwegian" or "German" denoted a linguistic community, but American immigrants in the nineteenth century could rarely
claim a unified national language as their mother tongue. Instead,
immigrants spoke a host of regional dialects that served as their
native languages. The "official" tongue of Norwegian Americans,
for instance, was often "Dano-Norwegian," a form of Danish that
emerged after Norway and Denmark united in the fourteenth century. But, multitudes of Dano-Norwegian dialects existed because of
regional and social-class variations. One commentator noted that in
Norway "every time you crossed a creek, there was a new language."
In addition to the Danish dialects, "New Norse," the brainchild of
linguist Ivar Aasen, captured the interest of many Norwegians in the
second half of the nineteenth century. The "new" language was actually a Norwegian folk language; it was hoped that New Norse would
bring the people back to their historic linguistic roots and set Norway
apart from the Danes and Swedes. Not all Norwegian Americans,

however, warmed up to the idea of using a folk language as a new mother tongue. In 1874, for instance, Norwegian immigrant Andreas Hjerpeland admonished his friend for his use of that "country language."[43]

The Germans in the United States also spoke a multitude of regional dialects. Because a dialect was an immigrant's *Muttersprache*, it was that language to which he or she was most loyal and hoped to cultivate among his or her children. Cincinnati's North German Lutheran Church, for instance, sought not so much to preserve the "German" language as it did its members' low-German dialect: *Plattdeutsch*. Low-German dialects, as well as some Scandinavian languages, were perhaps difficult to keep "pure" because of their linguistic similarity to English, and many low-German speakers picked up the American language with "startling rapidity." Pennsylvanian Dutch, which Boernstein described as "a jargon concocted of English and German words," was another common German dialect in the United States; it was the language of choice for several American newspapers. As the Forty-eighters came to the United States, they were disturbed by the condition of the German language in the new land. As Boernstein noted, "Among the great mass of educated immigrants, there was a desire and solid will to elevate German Americans intellectually [and linguistically] rather than to descend to their lower level." The political refugees, therefore, felt it was their mission to cultivate and preserve Germandom in the United States and to impose their academic and literary form of High German for the remainder of the German Americans, who spoke "common" German, "often with bad grammar." For the Forty-eighters and other German intellectuals, the foreign-language press was one vehicle for uniting the German Americans under the banner of Germandom and to spread their form of the German language. They were, in a sense, trying to create an imagined community among the diverse German element in the United States. Ethnic leaders, however, were not successful at creating a united ethnic community. Even as national organizations emerged to achieve ethnic solidarity—such as the German-American Alliance—they were largely able to rally only the support of like-minded members, such as the secular club Germans.[44]

Regardless of the difficulties involved in the endeavor, many ethnic leaders sought linguistic unity. The immigrants arrived in the United States after haven taken a tremendous risk; emigration was dangerous, arduous, and miserable. "But in the midst of this tidal wave of misery," novelist Kurt Vonnegut has written, "was what would in retrospect seem to the Anglos a Trojan horse, one filled with educated, well-fed, middle-class German businessmen and their families." These Germans, from which Vonnegut descended, established "their own banks and concert halls and social clubs and gymnasia and restaurants,...leaving the Anglos to wonder,...'Who the hell's country *is* this anyway?' "[45] It was both the immigrants' country and the Anglos'; America, in fact, had become a polyglot boardinghouse, a multilingual nation. And Vonnegut's ancestors in Indianapolis, along with scores of other ethnic leaders, tried to unify the diverse elements of their respective immigrant groups by setting up bilingual programs in the newly founded common schools.

2

Building the Polyglot Boardinghouse in the Northeast and the South

As the common schools emerged during the second third of the nineteenth century, almost nothing prevented some of them from becoming bilingual institutions. The American tradition of localism and, with it, control over community institutions facilitated public bilingual schooling. So too did the nation's enormous land holdings; groups of linguistically similar folks could simply establish a town and its common schools would reflect the language of the community. The foreign-language-speaking immigrants also were anxious about the America's language. Ethnics, therefore, used bilingual education to subdue some of their nervousness by simultaneously promoting both the language of their home and the language of their new homeland. Some nineteenth-century schools became dual-language institutions naturally because of the populations they served, while others developed as bilingual as a way of enticing foreign-language speakers into the public sector. In addition, some common schools emerged as bilingual institutions because of the demands of politically powerful immigrant groups within the community, groups that insisted that their perceived superior culture and educational practices be included in the public school curriculum. Occasionally schools transformed into bilingual institutions as a political appeasement, while some schools, although officially opposed to bilingual education, utilized dual-language instruction in a covert manner.

While immigrants felt a sense of anxiety about the English language, they also feared language loss among their children. One Norwegian immigrant noted to her family that her "Lillegut," who recently began his schooling, "speaks English and Norwegian on top of each other. You would laugh if you heard him." When faced with their children's failing command of the mother tongue, however, other immigrants did not take the situation so lightly. In 1850, for instance, Wilhelm Krumme was upset that his eleven-year-old son "doesn't know much German" and hoped, if the family situation improved, that the boy could be taught the language. Adolf Douai, a prominent German educator in New York, captured the sentiment of many German immigrants. "The German-born population[s]," Douai wrote in 1870, "find that their children rapidly unlearn the German tongue, [which]....disturbs the family relations, the efforts of parents toward the education of their children, and the respect due to the parents from the latter." In the American environment, it was difficult to keep the mother tongue "pure" and free of the "Anglicisms that had contaminated it." The immigrant "youth," one newcomer lamented, "suffers temptations from the English. Many are nothing but heathens." Among the many temptations was the exclusive use of the English language, the language of the common schools. Henry Boernstein, for instance, noted that the "youths, growing up in the free schools, were either Americanized or rendered entirely savage." The situation had become so bad that, as one leading German American in Indianapolis, Theodore Stempfel, complained, "conversing in English" had become commonplace at "events of the German societies" after the Civil War.[1] Bilingual schooling, therefore, was considered a partial remedy for this anxiety over language loss.

But, public bilingual education did not take root in all regions equally. Constructing a polyglot boardinghouse that was sustained by dual-language instruction proved difficult in some areas of the nation. Except for a few locations, the South did not have major populations of foreign-language speakers. Although common schools emerged in some of the more populated coastal cities during the antebellum period—cities such as Charleston, Mobile, Savannah, and, of course, New Orleans—the South, in general, did not build its systems of public education until after the Civil War. The Northeast too was not a stronghold of public bilingual education. While the Northeast was a major destination of many foreign-language-speaking immigrants, its communities and public schools were often the makings of native-born citizens, not of the newcomers. Although not the most favorable of environments, the South and the Northeast did witness the rise of public bilingual schooling prior to the 1880s, but, because dual-language instruction was not the fundamental focus of public schools, the historical literature and primary sources for these regions are sparse compared to other areas of the nation. In fact, it was the

Midwest, the focus of the next chapter, that became of the hub for bilingual education. In the emerging towns of America's heartland, Heinz Kloss has insightfully noted, the immigrants did "not look upon themselves as guests in a well-established commonwealth" as in the Northeast and South, but saw themselves "as co-founders and partners in a newly-founded enterprise."[2]

Bilingual education did not take root among all linguistic groups equally either. In the Northeast and South, it was primarily the German-speaking immigrants who were most successful in setting up bilingual programs in the emerging public schools. Kloss, the late sociolinguist and onetime National Socialist, has argued that in the mid-nineteenth century "the Germans were the only important immigrant group." This assertion, although seemingly contaminated with the stain of Kloss' Nazi past, contained a certain amount of truth. Germans certainly were not the "only important" ethnic group in the United States, but they were numerically large (see table 1), which was part of Kloss' point. In fact, Kloss has estimated that by the early-twentieth century there were nearly nine million German-language speakers in America, which is a relatively conservative estimate because it does not seem to account for the third generation immigrants who also spoke some form of the language.[3] Nevertheless, the Germans' numerical strength in the United States contributed to their success at implementing bilingual programs in American public schools.

In addition to being a numerically "important" group, the Germans were also a favored ethnic group at times during the nineteenth century. In general, Germans were perceived by Americans as honest and hard-working people. Even American textbooks during the century portrayed them "in a most favorable light." William T. Harris, the superintendent of the St. Louis schools, described the Germans as intensely thoughtful and among "the most peaceful and law-abiding citizens in the world." Contributing to this positive perception of the Germans were the many American intellectuals and educational leaders who were Germanophiles, including Harris, the Hegelian "philosopher-king of St. Louis." Beginning in the early-nineteenth century, several American scholars made their way to the German lands to receive advanced graduate training, which was relatively unknown in the United States. These intellectuals, including George Ticknor of Harvard, were so taken with the specialization, academic freedom, and quality of research they encountered overseas that they hoped to bring the "German model" back to America. They looked to the German universities for an archetype of collegiate reform in the

United States. It was not until the latter part of the century, however, that the German university model would make significant inroads on American soil. The common school reformers also looked to the German states for educational inspiration. For instance, Calvin Stowe of Ohio—husband of the celebrated author Harriet Beecher Stowe— was enamored with the Prussian schools after a tour of Europe. In 1837, Stowe praised the "practical character of the system" of education in Prussia, making special note of how the eastern Germans were able to seamlessly insert moral and religious study into the curriculum. Stowe also commented on the Europeans' success with bilingual instruction. "Two languages," Stowe came to believe, "can be taught in a school quite as easily as one, provided the teacher be perfectly familiar" with both.[4] With so many important champions of Germans and their educational systems, the German language easily found a home in many public schools across the nation.

While America's general Germanophilia facilitated the favored status of German in the public schools, nothing contributed more to the emergence of German-English bilingual programs than the arrival of that "Trojan horse" that Kurt Vonnegut has described, the one filled with intelligent Germans who were more than willing to take an active role in reshaping American life. Perhaps the most notable of those in the Trojan horse were the Forty-eighters. "Neither America nor any other land," gushed the Forty-eighter Boernstein, "has ever experienced an immigration such as the German of 1848, and this grandiose phenomenon will certainly never be repeated." What made this immigration so unique was that "a mass of educated minds, gifted disciples of the sciences and the arts, tested pedagogues and shapers of youth came to America in the space of a few years." In short, "a plethora of knowledge, learning, intelligence and ability was poured out on the whole Union!" These Forty-eighters, with their moustaches, beards, and general "bohemian appearance," were devoted to raising the intellectual level of German Americans. German-language periodicals, low-cost classics in German literature, and theater troupes were only a few of the educational and cultural endeavors that were made widely available by the political refugees. Yet for these elite immigrants, Germany was a "Kulturnation," a land that was best seen as "a cultural ideal rather than as a political construct." The aim, therefore, was to ensure that the Fatherland's culture and knowledge did not dissipate in the new American setting, to ensure, in fact, that the Kulturnation thrived on American soil. These bohemian-looking refugees were determined to provide the *Weltanschauung* for the German Americans, and the emerging public schools proved a useful tool for that endeavor. The Forty-eighters' sentiments were vividly captured by a drama written by Dr. Rudolph Guszmann, a friend of Boernstein. In the opening monologue, Guszmann told the St. Louis audience in January 1854 that "Even if the hatred of princes has expelled you, Robbed you of your homeland, yours, beloved, German, You still have German art and science to call your own! Let it propagate now on free soil! Let free schools be its protectors!"[5]

While the numerical strength of the Germans, America's German-ophilia, and the arrival of the highly educated Forty-eighters all contributed to the special status that German was to have in the public schools of the United States, other languages found their way into the common schools as well. In New York, for instance, some Native American languages were fostered in a few public schools, as was French. The French language was also cultivated in New England and Louisiana, particularly in New Orleans, and in a few locations in between. But, it was German that dominated the bilingual common school curriculum in the Northeast and the South prior to the Progressive era.

Polyglot Schooling in the Northeast

In many respects, New York City was the center of immigrant activity during the nineteenth century. It was both a central destination for foreign-language speakers and a major receiving port from Europe. With thousands of immigrants arriving weekly, in August 1855 the city opened Castle Garden to process the newcomers on the southern shores of Manhattan; the former fort was replaced by the newly constructed Ellis Island in the 1890s. Although New York City was tremendously multilingual, its common schools, for the most part, did not reflect this linguistic diversity in their official curricula. Public schooling emerged in the city as the New York Free School Society consolidated many of the city's charity schools into a coherent system. By the mid-1820s, the Free School Society became the Public School Society and sought to attract not only the poor children, as the charity schools did, but all the city's youngsters in order to provide them with a common education. By the late 1820s, the Public School Society's schools were "common schools," financed by taxation. In 1853, the Society's educational duties became the responsibility of the newly created Board of Education of the City of New York. In the process of developing a public school system, Manhattan had taken to heart the common school ideology of training the students in the habits of "self-control" in order to create a more homogenous and, therefore, harmonious society.[6] The common schools of New York and their aims were, thus, already in place as the massive waves of foreign-language speakers washed up onto the American shores.

In 1854, the German language was introduced in the last grade of New York City's grammar schools. Not bilingual in intent, this one-year, optional program in the seventh grade added a modern language elective to the traditional studies of reading, spelling, arithmetic, and geography. By 1870, however, German became a part of the regular elementary school curriculum "because of the increasing importance of the German element" in the city. For the next six years, the language was offered in all of the elementary grades; in 1873, nearly 20,000 students enrolled in the city's 464 German classes. Manhattan's limited German-language program was not aimed at producing bilingual students, but "as a means of Americanizing the citizens of foreign birth and their children." The fear was that German children were shying away from the public schools in order to attend German private and parochial schools. In fact, many were. During the 1868–1869 school year, for instance, New York City had 350 private schools, which enrolled 50,000 children. In the city there were several private secular German schools, including the New York German Free School and Douai's two private German schools, one for each sex. Therefore, the goal of the German classes, as one school commissioner noted, was to attract the immigrants to the public schools in order to bring "them in immediate and daily contact with the children of Anglo-American birth, and under such educational and moral influence which tends to make them a thoroughly homogeneous part of our nation."[7] The plan did not seem to be all that successful. In 1874, the commissioner stated that "at least 11,000 German pupils are in daily attendance at the Catholic parochial, Lutheran, and German private schools." During the 1875–1876 school year, New York could still claim 350 private schools, but their enrollment had increased by ten thousand. Perhaps partially because the German program did not seem to be putting the private and parochial schools out of business, its position in the public schools of New York was quite precarious. After much debate in 1875, the German-language program was once again pushed back into the higher levels of the grammar schools for the following school year, although this time German classes began in the fifth grade. (There were six primary grades and seven grammar grades in the city's schools of the 1870s). In New York City, the

German language would remain in the grammar schools until the Progressive era.[8]

Several of the normal schools of the Empire State supplied the teachers of German for Manhattan's public schools. These future educators learned the methods of the day for teaching a foreign language, which often focused on grammar, reading, and translation. By the 1860s, however, the "natural method" gained a modicum of popularity in the United States. This method, along with the "direct method"—the more-moderate cousin that became prevalent after the 1880s—shifted the focus away from grammar and translation to comprehension and conversation. In the 1870s, for instance, the State Normal and Training School at Brockport reported that translation had become of "secondary" importance in its foreign-language curriculum. Although the new methods were changing the teaching of foreign languages, the traditional "grammar-translation" method persisted. Even at the turn of the twentieth century, to get a license to teach German in New York City's public schools required the candidate to translate German literature—often Schiller and Goethe—into English during the three-hour licensing examination. Enrollment in the normal schools' German classes was high in New York. In 1874, for example, out of 1,200 students only twenty did not study the German language at the Female Normal College. French too was offered at some of the normal schools, since the language was also an elective subject in the higher grades of the city's grammar schools. Yet, French was not nearly as attractive to the future teachers as German. At the Female Normal College, only twenty students enrolled in the French program. At some normal schools, however, *français* was exclusively offered. At the State Normal and Training School at Potsdam, for instance, French was the only modern language taught, which was not surprising given the large number of Acadians in upstate New York and farther northward. Although not as common as German in the nineteenth century, the French language was taught in several of New England's pre-secondary schools. Like German educators, the teachers of French were expected to be able to translate efficiently, a central skill needed to get a license for teaching *français* in New York City.[9]

New York City, of course, was not the only location in the Empire State that allowed for elementary foreign-language instruction and bilingual education. In 1866, Buffalo established an elective, elementary-level German-language program in its public schools, an initiative that had been considered since the late 1830s. In the early

1870s, the western city in New York offered the language in fourteen schools, and nearly nine hundred students took advantage of the opportunity. The introduction of German in Buffalo's public schools was largely a political maneuver. Prior to the Civil War, the German immigrants tended to cast their ballots for the more pluralistic Democrats and shied away from the nativist Whigs. After the demise of the Whig-oriented, Know-Nothing nativism and the rise of the Republican Party—a party that paralleled many of the humanitarian values of the more liberal Germans, particularly its stance against slavery—some immigrants increasingly became attracted to the new party. After the Civil War, as historian David A. Gerber has argued, the German vote was openly available, and neither party wanted to antagonize such a numerically powerful segment of the population. In addition, having fought together in the war, tolerance characterized the relations between the native born and the immigrant during the1860s and 1870s, unlike the unbridled nativism of the 1850s. With a generally relaxed attitude regarding ethnicity and a large number of ballots potentially going to either the Democrats or the Republicans, political accommodations were inevitable, including the introduction of the German language into the public schools. In short, the German-language program in western New York was primarily a political tool to appease the immigrant population.[10]

While Buffalo introduced German into its elementary schools, so too did Brooklyn and Queens, which on January 1, 1898 became part of New York City. But, German and French were not the only non-English tongues spoken in state. The State of New York controlled eight reservations, which, in the 1870s, contained over five thousand American Indians. Although members of twelve tribes inhabited New York, the vast majority of the Native Americans in the state was Seneca, part of the Iroquois Confederation. In 1868, twenty-six schools served the nearly two thousand children on these reservations; less than half attended, however. As with the immigrants in Manhattan, the goal of the reservation schools was not to promote bilingual or bicultural education, but to "civilize and christianize this singular race of men and fit them for citizenship."[11] By the 1870s, several Native Americans had joined the teaching staff on the Cattaraugus reservation. Although the Americanization focus of the schools did not change, these Indian teachers likely educated their students bilingually, at least in informal (and potentially subversive) ways when their pupils failed to understand the English language, as was the case at other Native American schools. In addition, the Onondaga reservation was partly served by missionaries, and many missionaries in the United States were known for their "bilingual approach" to Christianizing the American Indians. Jonathan Kneeland, the superintendent of schools on the Onondaga reservation, admired the missionaries who "gained their [the Indians'] good will and spoke their language fluently" but did not openly admit to the state superintendent that bilingual education was used in his schools.[12]

South of New York, bilingual education also developed, particularly in the mid-Atlantic states of New Jersey, Pennsylvania, and Maryland. In the 1860s, Newark permitted German in its primary schools as long as the teacher also was fluent in English. As a region, the mid-Atlantic was a major destination of the waves of German immigrants arriving during the nineteenth century; only the Midwest received more Germanic settlers. Yet, these were not only heavily Teutonic states, they had also been Germanic colonies. Many of the German speakers who arrived in the New World during the eighteenth century had planted their roots in the middle colonies, colonies known for their tolerance. The German colonists were a mixed lot, representing, as one traveler in New Jersey and Pennsylvania noted in 1734, at least sixteen different religious sects and faiths.[13] With a strong Germanic presence in the region for over a century, many of the common schools in the mid-Atlantic emerged naturally as bilingual institutions.

Although there had long been state laws in Pennsylvania that funded schools for poor children at the public's expense, the first common school bill—guaranteeing public schools for *all* children—was passed in 1834. (Philadelphia had established its own school system in 1818.) Drafted by State Senator Dr. Samuel Breck—a Philadelphian who, as a friend described him, "was a gentleman of fortune, without children, and with a heart moved by feelings of the warmest philanthropy and benevolence"—the Pennsylvania school law "in practice...proved defective in many respects." The law, therefore, was revised several times over the subsequent years. In the late 1830s, for instance, the common school legislation recognized German as one of the languages of instruction in the state's public schools. This linguistic addition to the state's public education law was inspired by Stowe's report on the Prussian schools. Some of most sweeping changes in the Pennsylvania school law came in 1854, when a system of county superintendence was put into place. Although the creation of state and county-level superintendents was certainly a move toward centralization, the revised legislation ensured that localism in the realm of education would still reign supreme. Local officials, the law stated, would "determine and direct what branches of learning should be taught."[14]

Local school leaders would also determine the language in which those branches would be taught, and in many locations throughout the state the language of instruction, at least partially, was German. In Berks County, where—as County Superintendent William A. Good noted in 1857—"the German element predominates largely," the recommended English textbooks were not very successful. The "pupils," Good reported to the state superintendent, "passed over much . . . without a knowledge of the very elements of the language of those books." The county superintendent recommended that many of the schools remain bilingual, thus ensuring that "the rising generation will grow up acquainted with the English as well as the German language," which, he added, "must prove advantageous." German was a regular subject in many of the public elementary schools throughout the state. Over half of Pennsylvania's counties contained students studying the language during the 1850s and 1860s. In 1862, for instance, the German language was taught in 69 percent of Pennsylvania's sixty-four counties. Numerically, Berks and Lehigh had the most students; both counties had over 1,000 pupils learning the language throughout the 1860s. In fact, between 65 and 76 percent of the districts in the two counties taught German in the 1850s and 1860s, and in some districts over a third of the students enrolled in German classes. Elk County, although smaller than both Berks and Lehigh, had a higher percentage of students studying the language. However, the use of German, particularly among the younger generations, gradually began to decline in all the Pennsylvania counties after 1860, a trend that Stempfel also witnessed in Indianapolis (see table 2 and table 3).[15]

German, as the language of instruction or as a subject, was a necessity in many of the counties in Pennsylvania not only because the students themselves were native German speakers, but also because the success of the common schools depended upon it. Many of the Germans in the state were rural people who were not yet convinced of the value of public education, and they often feared that their children would become Americanized in such schools. J. S. Ermentrout, the Berks county superintendent in the 1860s, understood that apprehension and fear. "[O]ur people," the county superintendent noted, "once convinced that, in being educated, they will not be called upon to ignore their original [German] character, the last vestige [sic] of

Table 2   Pennsylvania German Students in 1857[i]

County	Total Number of Students	Number of German Students	Percentage of German Students
Berks	22,179	1,726	8
Elk	1,028	300	29
Lehigh	9,644	1,469	15

Note:

i. The total number of students represents the combined number of male and female students, not those in daily attendance. Using the daily attendance yields a much higher percent of German students; in Elk County, for instance, 39 percent of the students enrolled in German classes.

Source: Pennsylvania Common Schools, *Report of the Superintendent of Common Schools of Pennsylvania, for the Year Ending June 2, 1857* (Harrisburg: A. Boyd Hamilton, 1858), 132–133.

Table 3   Pennsylvania German Students in 1861

County	Total Number of Students	Number of German Students	Percentage of German Students
Berks	24,350	1,475	6
Elk	1,209	288	24
Lehigh	12,210	1,057	9

Source: Pennsylvania Common Schools, *Report of the Superintendent of Common Schools of Pennsylvania, for the Year Ending June 3, 1861* (Harrisburg: A. Boyd Hamilton, 1862), 252–253.

opposition to the public schools will have disappeared." Keeping their original character meant retaining the mother tongue in the schools, as well as—as some schools did—using the German Bible as a text. Native Germans were employed in the schools as teachers. They, as Ermentrout noted in 1863, were among the best prepared educators. Also to convince the local populations of the importance of public schooling, one county superintendent recommended that educational journals be published in German. While learning the language of the land was important, Ermentrout claimed, "in order to learn English, it is not necessary—as some would have us believe—to ignore the German, and wage a war of extermination against the customs and modes of thought that characterize the German counties of the State."[16]

The German-English public schools in Baltimore were not an attempt at Americanizing the immigrants, as in New York City, nor did they become bilingual naturally, as in the rural districts throughout Pennsylvania. Instead, the bilingual schools emerged in the 1870s as the city's school board recognized not only "the value of the language in the business relations of society," but also that "many of our citizens are of German origin, whose children need and desire the educational advantages of our schools, but cannot obtain them in consequence of the absence of their language."[17] As a major port of entry for immigrants, Baltimore—a city on the Mason-Dixon line, but with an immigration pattern more aligned with the North than the South—became a heavily Germanic urban area during the nineteenth century. By the latter part of the century, Germans made up nearly a quarter of the 434,000 inhabitants. Many of the Germans in the city, both church and club Germans, sent their children to private schools where the mother tongue was utilized. Private schooling was common among the German Americans in the nineteenth century because it not only maintained the religious and linguistic traditions of the immigrants, but also, especially for the German intellectuals, because it could make use of the progressive methods found in the schools of Germany. The still-developing public schools of the United States were seen by many of the "better class of Germans"—the Forty-eighters and their less-radical allies, the Thirtiers—as intellectually, linguistically, and pedagogically inferior. The Baltimore Germans, therefore, relied heavily on private schools for educating their children during the first three quarters of the century. The most noteworthy German private schools were Heinrich Scheib's Lutheran Zion School, Friedrich Knapp's German and English Institute, and Philipp Wacker's German-English School; the latter two schools were secular—Knapp, in fact, was a Forty-eighter—and employed Pestalozzian techniques.[18]

In 1873, the school board and other political leaders in Baltimore saw the need for public schools that offered German, and, as in Buffalo, the Germans represented an important voting bloc that needed to be conciliated. But, German as an elective subject, as in New York, would certainly not motivate the immigrants to send their children to the city's common schools. The public schools had to be pedagogically progressive, allow for dual-language instruction, and introduce the "treasures of the German literature" in order to find favor among middle-class Germans and intellectuals. During the subsequent years, the Baltimore public schools did just that; by 1876, the city operated five fully bilingual schools, one of which had been Wacker's German-English School. The German-English schools offered both primary and grammar-level instruction, and both German and English-speaking children attended the dual-language schools. These schools became quite popular, so popular in fact that the secular private schools declined steadily after the opening of the bilingual public institutions. In 1876, the five schools enrolled nearly three thousand pupils, and the number of students and schools continued to climb for the remainder of the century. The German-English schools, although technically controlled by the Baltimore Board of Commissioners, were operated and guided by German-Americans liberals such as Karl Hessler. While the Germans were able to see most of their educational demands realized in the German-English schools, physical education, long an aim of the Turners, was not introduced into the public schools until 1895. Once the subject was introduced, the German-American Turners, naturally, supervised that endeavor.[19]

Polyglot Schooling in the South

Although the South—that region east of the Mississippi and below the Mason-Dixon line—lagged behind the Northeast in the development of common schools, several port cities, located along the Atlantic Coast, Gulf Coast, and Ohio River, did establish formidable public schools systems relatively early. Ports attracted inhabitants, many of whom were foreign-language speakers. Carl Berthold, an immigrant from the German principality of Waldeck, found "Louiswille...a beautiful city." Louisville, Berthold told his family in 1854, "lies high on the banks of the Oheuo... [and] has good drinking water; the number of inhabitants is 60,000, of which 20,000 are Germans[,] the rest mostly natives." Berthold also claimed that a "tenth of Louisw. is now made up of Waldeckers," which was surely an exaggeration, as was his report that a third of the town was Germanic; in 1880, Germans made up only 11 percent of the population. Perhaps one reason why Berthold was so enamored with Louisville was because the city was home to many like-minded Germans. Berthold was an educated, liberal Forty-eighter, one who had been active in the Turner societies back in the Old Country. As in Baltimore, the liberal club Germans were a dominant force in Louisville, a town also on the border of the Mason-Dixon line. It was, in fact, the city in which the notorious "Louisville Platform" was written. In the mid-1850s, Karl Heinzen and other freethinking liberals insisted that women "shall have equal rights with men" and demanded an end to the "political and moral cancer" of slavery. Perhaps less controversial, the Louisville Germans also called for "free schools, and German teachers where there are German settlements."[20]

In much of Kentucky, the common schools emerged only slowly. Fraught with financial difficulties, the very existence of public schooling in the state was rather precarious up until the 1870s. In Louisville, by contrast, common schools originated in the second quarter of the nineteenth century, as in the more-developed areas of the Northeast. Although many of the state's schools struggled for survival in 1870, the Louisville schools in the same year were remarkable. By the end of the 1860s, the city could claim fourteen "ward" schools and two high schools, and these schools enrolled more than twelve thousand pupils. The ward schools consisted of ten grades that spanned the primary, intermediate, and grammar levels, and the German language was offered at all levels, in grades one through nine. The elementary students had a beginning German course, one that focused on the rudiments of reading and writing the language. By the grammar grades, the German students continued their reading, writing, and spelling lessons, but also began to translate German materials into English. Louisville's two high schools, one for each sex, allowed students to develop their understanding of German even more in their two-year program, although, at the secondary level, students had the option of learning French as well. In a city dominated by that "better class of Germans"—the liberal club Germans who demanded language maintenance and progressive education—it was no surprise that the German language, as well as "[n]ew and natural methods of instruction," found their way into the Louisville public schools. It was also no surprise that several of the city's Germanic inhabitants came to hold important positions within the school system, including positions on the school board.[21]

While bilingual education took root on the banks of the Ohio, its presence was even more pronounced at the mouth of the Mississippi. After President Jefferson's Louisiana Purchase in 1803, more and more Anglos made their way south to the largely French city of New Orleans. As schools began to emerge in the 1820s, the Crescent City natives—the French-speaking Creoles—entered into a compromise with the English-speaking newcomers. The city would have two primary schools. The downtown primary school and the secondary school, *L'Ecole Central*, were controlled by the Creoles, while the Anglos operated the uptown primary school. All these schools became bilingual naturally, out of necessity because of the linguistic needs of the population. This downtown-uptown arrangement did not last, however. By the late 1830s, New Orleans reorganized its political structure; the city was divided into three municipalities, divided, primarily, along linguistic lines. The first and third municipalities consisted primarily of French Creoles, while the second municipality was essentially Anglo. Each municipality, of course, controlled its own educational affairs. Joshua Baldwin of the second municipality wanted to reform the schools in his section of the city. He greatly admired the work Horace Mann was doing with the common schools of Massachusetts and wrote to him for assistance. In the early 1840s, Mann sent Baldwin a New Englander, John Shaw, to help reorganize the schools along the lines of those in the Bay State. The second municipality, therefore, came to share the educational ideology of New England. By contrast, the first municipality, encompassing the French-speaking downtown area, clung to its more traditional school arrangements, including bilingual education.[22]

As the first and third municipalities witnessed the successes in the schools of the second district, they emulated many of the reforms, although bilingual education still persisted. By the 1840s, however, the first municipality experienced a demographic shift; numerous Irish and German immigrants settled in the downtown area. With more non-French-speaking newcomers inhabiting the French Quarter, the municipality's bilingual schooling came under attack. In 1843, the non-French elements carried the municipal elections and quickly ended the bilingual programs, although *français* remained a branch of study in the schools. The new educational leader for the municipality, Christian Roselius, saw the virtues of the New England-inspired educational system in the neighboring second municipality and sought out its assistance for reforming the schools in his district. Roselius, a stern-looking German immigrant, apparently did not see the irony in his work to end the bilingualism of the French Quarter, for it was bilingualism that his fellow countrymen were attempting to preserve in other parts of the nation. The following year yielded a Creole victory in the municipal elections, and the French-speakers reinstated the bilingual programs. This back and forth power struggle between the Creoles and newcomers continued for several years. Eventually, however, many of the educational reforms of the non-French elements rooted themselves in the first municipality. The grading system, for instance, took hold in the schools, and two high schools, one for each sex, were established; these secondary schools supplied the bilingual teachers for the elementary schools. In 1853, for the first time the combined number of non-French-speaking students surpassed those with a French mother tongue in the first municipality (then called the second district), although French students still comprised the largest group.[23]

The Civil War, however, greatly altered the schools of New Orleans. The war, in fact, altered more than the schools; it changed the worldview of Southerners. How could it not when, as Mark Twain discovered, "every man you meet was in the war; and every lady you meet saw the war." In the South, Twain commented, "the war is what A.D. is elsewhere; they date from it." Benjamin F. Butler—the slightly plump, bald-headed Union general who wore a long mustache—ushered his troops into New Orleans in 1862 and took control of the city, including its schools. With the Yankees in power, General Butler, and later General Nathanial Banks, essentially imposed a northern educational model on the Crescent City, a model that demanded the centralization of the public schools. The new consolidated educational system was intended to promote fidelity to the Union cause, and, therefore, teachers had to sign loyalty oaths. The new educational mission and arrangement—which, based on New England common-school principles, were to create a homogeneous citizenry—undermined the bilingual schooling of the French Quarter. Homogeneity meant English would be the language of instruction, and the Yankees, therefore, ended the bilingual education in the city's primary schools. French was retained as a subject in the higher grades, but only as a means of appeasing the local population in order to ensure that children still attended the public schools. After the war, the Confederate Southerners regained control of the city's affairs and hoped to undo much of what the Yankees had accomplished. (The loyalty to the Confederate cause was so strong in New Orleans that "Lee Day" came to be an official school holiday.) The Southerners, however, kept French as merely a branch of study, although it was allowed in the primary schools of the French Quarter.[24] Once English made significant inroads in the curriculum for a number of years, it was difficult to return to bilingual education.

While German was the dominant non-English language in the common schools, it, along with other languages, did not become part of the curriculum in the same manner or for the same purposes in every locality that offered foreign-language study. In fact, the common schools of the Northeast and South brought forth distinct patterns of bilingual education and foreign-language instruction. In some areas, as in New York City, the addition of German to the curriculum was an attempt to attract immigrants to the public schools in order to assimilate them. After the Union takeover of New Orleans, the study of French was used in a similar fashion. In other areas, token foreign-language instruction was a politically motivated endeavor used to appease a large segment of the population. In the rural schools of Pennsylvania and in antebellum New Orleans, public bilingual schooling was a natural endeavor, emerging simply because of the large number of non-English-speaking children. In such enclaves, bilingual education was a necessity if common schools were to take hold. In Baltimore and Louisville, bilingual education developed not as a natural undertaking, but because of the demands and needs of politically powerful liberal Germans who wanted their modes of life and cultural values reflected in the curriculum. On the Native American reservations in New York, bilingual instruction was largely a covert activity, one that facilitated students' understanding of the English curriculum. Some of these patterns were repeated—but with even more intensity—in the parlor of the boardinghouse: the Midwest.

3

Inside the Boardinghouse's Parlor

The parlor emerged as the primary space of middle-class homes in Victorian America. It was the focal point of the house, a room so prominent that—because of the architectural style of the time that positioned the room as a separate wing—it could often be detected from outdoors. The parlor was the public area of the home, and it was often used a cultural center, a gathering place that showcased the owner's literary and artistic tastes. Like the Victorian parlor, the Midwest was the centerpiece of the polyglot boardinghouse. The area that spanned along and east of the Mississippi and north of the Mason-Dixon line was where public bilingual schooling planted its roots deep into the American soil and proliferated before the 1880s. Bilingual education took hold in the Midwestern schools not only because of the enormous number of foreign-language speakers who settled there, but also because it was a developing area when the immigrants arrived, thus allowing them to become "co-founders and partners" in the region's affairs.[1] The immigrants did not shy away from the Midwestern parlor. They too entered into the public sphere, particularly in the realm of education, and displayed their cultural preferences.

Many writers who were intimately familiar with the nineteenth-century Midwest expounded interesting characterizations of the region and the nation to which it belonged. Several commentators were struck by America's "vigor." Mark Twain, having grown up in Missouri, stated that the Americans of the nineteenth century "don't dream; they work." America's vigorous work ethic was often attributed to the country's youth. Henry Boernstein, the famed German editor from St. Louis who also once lived east of the Mississippi in Illinois, wrote, "Old people will die here like everywhere;...this is no country for old folks—they must give way to the young ones, who are to work and improve; a young country wants young powers." Novelist Theodore Dreiser, who came of age in nineteenth-century Indiana, agreed that the United States "is a mere child as yet" but was occasionally pessimistic about that stage. Although "youth" brought "vigor" and "courage," it also meant that the nation's "problems are all before" it.[2] If the country itself was young and hard-working, the Midwest was even more so. Still in its infancy, but a bundle of developing energy, the heartland of America was the ideal environment for the establishment of public bilingual education.

The nineteenth-century Midwest was unique. On his return from a visit to his native Indiana, Dreiser concluded that the United States was "divided into distinct" regions; the Midwest, especially the Hoosier state, was "a world all unto itself." The Midwest was a "tobacco-chewing region," as Twain characterized it in the 1870s, a region where the men still wore boots, goatees, and mustaches, fashions long out-of-style in the East. America's heartland was largely rural, an area where in the evenings one could hear "the calls of farmers after pigs, the mooing of cows, the rasping of guinea hens, and the last faint twitterings of birds." There was also the "fragrant" aroma of the "soil and trees." The "geography" of the region, Kurt Vonnegut has written, inspired "awe for a fertile continent stretching forever in all directions. Makes you religious. Takes your breath away." The Midwesterner, like the American in general, felt free in his wide-open lands, where "[t]here are no ruling classes to him." Localism ruled in the heartland of the United States. Local "life," wrote the German-American Dreiser, "surrounds one like a sea. We swim in it, whether we will or no." Although largely rural, the Midwest was also dotted with emerging urban areas, and cities became the principal locations of educational innovation during the nineteenth century.[3] With a strong tradition of localism and an enormous immigrant population, the Midwest's cities, such as Chicago, Cincinnati, Cleveland, Indianapolis, Milwaukee, and St. Louis, emerged as the centers of public bilingual schooling in nineteenth-century America.

The Midwest's unique environment, where foreign-language speakers were "partners" in the development of the region's institutions and political structure, facilitated the emergence of numerous state school laws that actually encouraged bilingual instruction in some localities. While Pennsylvania, with its long history of German settlers in the state, was the first to pass legislation that allowed for German public schools in 1837, the nation's heartland was the primary location for this type of state-level activity. The Buckeye State led the way. The Germans of Ohio, especially those in Cincinnati, lobbied for an amendment that would legalize their plans for German instruction in the public schools, and the legislation passed in 1839. In 1869, Indiana too established a law that allowed for German in the public schools. Introduced by State Representative John R. Coffroth of Huntington, the German legislation had multiple aims. The Hoosier law was partially promoted as an economic move, an attempt to entice more hard-working Germans—that favored immigrant group—into the state. As in Ohio, the law also was billed as an attempt to alleviate some of the financial burden of the German settlers' educational costs. In addition to local property taxes, many German Americans, especially those in Indianapolis, paid private tuition fees so that their children could benefit from German pedagogical techniques and language instruction in parochial and independent schools. After the legislation passed almost unanimously, the amended school law in Indiana stated that when the parents of twenty-five students made the request, "it shall be the duty of the School Trustee or Trustees of said township, town or city, to procure efficient teachers, and introduce the German language, as a branch of study, into such schools; and the tuition in said schools shall be without charge."[4]

Indiana's law was typical; many of the Midwestern school laws that allowed for foreign-language instruction read similarly, although some states did not specify German. The states in the upper Midwest—Wisconsin (1854), Illinois (1857), Iowa (1861), and Minnesota (1867)—tended to allow for any foreign language, instead of just German, because of the presence of numerous Scandinavian settlers in the area. Some of state laws were amended over time, however. Illinois, Wisconsin, and Minnesota, for instance, all altered their original foreign-language laws at some point, but the legality of teaching any language in the schools of these states was retained. Ohio adjusted its schools law with regard to German in 1868 and 1873; the latter alteration stipulated that most of the school subjects be taught in English. Because schooling was largely a local endeavor prior to the Progressive era (and beyond), the state school laws merely made legal what was already occurring in some towns and cities. In some cases, however, the foreign-language legislation facilitated bilingual instruction in locations where common schools had not yet fully emerged.[5]

Although the state laws that fostered foreign-language instruction were an interesting feature of the Midwest, not all states in the region had such legislation on the books. Missouri, for instance, did not have a special school law allowing for bilingual instruction, although plenty of the schools in the state were German-English institutions. It was primarily local leaders who made decisions about the public schools' language policy, and at the local level many of the patterns found in the Northeast and South were present in the Midwest as well. That is, some localities used foreign-language instruction as a means of enticing immigrants into the public schools in order to assimilate them into America's culture and language. Other communities developed bilingual public schools naturally, simply because most of the residents were non-English speakers. Some areas, particularly in the larger cities, established bilingual programs to maintain immigrants' cultural and linguistic traditions, as well as to implement progressive educational techniques. A few localities offered token foreign-language instruction to appease a large and, therefore, politically significant ethnic group. As in the Northeast and South, the Germans led the way with regard to bilingual education. The Germans often thought the schools in the United States were substandard. As one immigrant noted in 1868, "I don't like the American school system.... The Americans think there's nothing in the world better than the schools here, which is a great self-deception. There are even many people here who cannot read a word." In the Midwest, it was primarily the secular Germans who utilized the developing public schools to maintain their cultural and linguistic values, while the church Germans made use of parochial schools for similar ends.[6] The Germans were not the only foreign-language-speaking group in the Midwest to develop public bilingual schools before the 1880s. The Norwegians of the upper Midwest also created a number of dual-language institutions. Like the region itself, the Norwegians generated a unique educational system, one that accommodated not only their linguistic needs, but also their desire for sectarian religious instruction.

Public Bilingual Schooling for the Creation of a "Better Class of Germans"

During the nineteenth century, the developing Midwestern cities were the focal point of bilingual education. In the 1860s, one German immigrant in Indiana told his brother that "for school teachers it's...better here than over there in Germany, in the big cities the teachers are well paid, from 30 to 50 dollars a month[;] if he knows German and

English he gets even more. [H]ere there's German and English schools and also ones where both are used." While the Midwestern cities were the hub of public bilingual education, it was Cincinnati—"the Athens of Western America"—that emerged as the model city for German-English schooling in the United States. Cincinnati's bilingual program became so well known that the city's German Department was "frequently called upon for information in regard to it." Common schooling began in the small river city in the 1830s. Like the urban Northeast, especially Boston, Cincinnati took public schooling seriously and became a beacon for educational improvement. In addition to being an educationally progressive city, Cincinnati was also a German city, and several German private schools opened simultaneously with the common schools. After Ohio revised its public-school law to allow for German instruction in 1839, Cincinnati began to use the language in some of its schools the following year. Originally, German instruction in the public schools was, at least officially, to facilitate students' transition to the English language. The German-English schools—two of them at first—were also promoted as an alternative to private schooling, which, because of its sectarian focus, did not appeal to all in the city's German community. The dual-language schools had a modest start, enrolling only a few hundred pupils the first couple of years. By the 1850s, however, the schools educated thousands of students bilingually, and gradually the aims of the German Department broadened.[7]

While the initial aim of German instruction was to ease immigrant students into an all-English curriculum, "[t]his," as Carolyn Toth has noted, "was not what the Cincinnati Germans had in mind," which was particularly true of the highly educated German refugees arriving in the 1840s and 1850s, as well as their middle-class, liberal allies who arrived a decade earlier. The liberal German element in the United States wanted, as Adolf Douai noted, "excellent schools," like those in Germany. The German educational system was "by far the best in the world" because the fatherland was "the cradle of the reformation of schools." This "better class of Germans," as Douai called those with a liberal outlook, also sought "the privilege of commanding the two master languages of the world, English and German, at the same time." German was important not only because it maintained the mother tongue, but also because the language was the key to "the treasures of the German literature," from which liberal German Americans found ethical inspiration and their worldview. Cincinnati's schools were already pedagogically progressive—employing such

Pestalozzian techniques as object lessons—which made the city's public schools even more attractive to the liberal German element.[8]

Although full of potential, the liberal German Americans in Cincinnati demanded more educational innovation. The German program, therefore, was reorganized in the 1850s to allow for the mother tongue in the grammar grades, which dramatically increased the enrollment in the German Department. The number of pupils jumped from just over sixteen hundred in 1855 to nearly five thousand in 1860. By the 1860s, the German Americans had a model school system. Gymnastics, long an aim of the Turners, was part of the regular curriculum, and, as the Committee on the German Department stated, "the two most beautiful of existing languages are brought together; the mother, the German, and the daughter, the English." The city's public schools, after much trial and error, established the "Cincinnati plan" of bilingual instruction during the 1860s. Half of each school day was devoted to German during the first four grades. The English and German teachers simply swapped classes, which added no additional expense to the school budget. Students wishing to continue their German studies in the grammar and secondary grades were given the option, but German at the higher levels of the educational system was only a branch of study, comprising about an hour of each school day. The Cincinnati plan proved exceedingly successful, garnering the support of the English-speaking superintendents such as John Peaslee, as well as the city's German community. By 1875, in fact, over fifteen thousand students—more than half of the city's pupils—received bilingual instruction. For their impressive program, the leaders of the German Department wrote their own bilingual textbooks, primarily because very few existed in the United States, but also because such books ensured that their progressive methods and worldview found their way into the German-English schools.[9]

Indianapolis, Milwaukee, and Cleveland followed the lead of progressive Cincinnati. In 1867, Clemens Vonnegut—a member of the Indianapolis school board and great grandfather of the novelist Kurt Vonnegut, Jr.—and A. C. Shortridge—the city's school superintendent—made their way to Cincinnati to study its bilingual schools, with the hope of improving their own German-language program, a program that was still in its rudimentary stages. By the time Vonnegut and Shortridge went to Ohio, in fact, common schooling in the Hoosier capital was not even fifteen years old. Public education had been a tough sell early on in Indianapolis. "Schooling," many Hoosiers believed, "led to extravagance and folly....A man could keep store, chop wood, physic, plow, plead and preach without an education, and what more was needed? .... Without the aid of Science, Nature had enriched us with the fruitfulest [sic] powers of mud. The wilderness of Indiana had been subdued, and teeming crops grew luxuriant over the graves of dead savages—all done by unlearned men." Although resistant at first, Indianapolis established public schools in the 1850s. By the 1860s—that decade during which the German vote was coveted by both Democrats and Republicans—the city had already begun to insert German-English education in a few of the districts where there was a high concentration of Teutonic citizens.[10]

Indianapolis, like many other Midwestern towns and cities, was heavily Germanic. A decade before the turn of the twentieth century, German immigrants and their kin comprised a third of the city's population. The leading Germans in Indianapolis, as in Cincinnati, were of the liberal type, and they were deeply involved with public education in the city. Vonnegut, for instance, was a freethinker, noting that he and his like-minded compatriots "base our faith on firm foundations, on Truth for putting into action our ideas which do not depend on fables and ideas which Science has long ago proven to be false." Charles Emmerich, Vonnegut's friend and fellow freethinker, became one of the most prominent supervisors of German in the city's public schools. These men, along with other club Germans, helped put a progressive stamp on the city. They also established German instruction in the public schools and, from their powerful positions within the schools, made sure their liberal values were reflected in the curriculum. Indianapolis's German program, however, got off to a slow start, employing only four teachers in the first few years. The city's German program was an optional branch of study that began in the second grade and continued through the high school years. During the 1870s and 1880s, the program really began to flourish. By the early 1880s, the German students in the upper levels of the grammar schools—grades six through eight—attended the "German annexes," which provided a half-German, half-English curriculum; even subject matter such as American history was taught in *Deutsch*.[11]

Milwaukee, like Cincinnati and Indianapolis, was a progressive city. By the early-twentieth century, local affairs, including public education, would be in the hands of the Social-Democratic Party. Milwaukee was also heavily Teutonic. In the 1850s, for example, over a third of the city's population was German. Writing from Milwaukee in 1847, Johann Pritzlaff reported to his brother that "you could set up a school in the town, for there are very many Germans living here.

Up until now there has always been a lack of decent schoolteachers."
Pritzlaff was a church German, and many parochial schools dotted
Milwaukee. However, many Germans, particularly those with secular
leanings, had their eyes on the emerging common schools, and they—
that "better class of Germans"—shaped the city's public educational
system. By the late 1860s, these liberal Germans, many of whom were
Turners and freethinkers, made up half of the school board and introduced the German language in the common schools; by the 1870s,
the vast majority of the school board members had Germanic origins.
German was an optional branch of study with short daily lessons,
and, originally, the English and German-speaking children learned
the language together. Although Milwaukee never developed half-
English, half-German schools as in Cincinnati and Indianapolis—
Wisconsin limited foreign-language instruction to one hour daily in
1869—the German program was revised to better serve the linguistic
needs of the pupils. In 1873, for instance, native-English and native-
German speakers were put into separate classes, an initiative that was
driven by pedagogical and language-acquisition concerns. In addition, German became a mandatory subject in the heavily Germanic
districts. Milwaukee's German program proved attractive to the city's
immigrants. In 1876, over half of the pupils in the public schools
studied the language, nearly 80 percent of whom "were of German
parentage."[12]

Cleveland too was a Germanic city whose public schools were
shaped by liberal-minded Germans. The push for German instruction
initially came from the school board, a move that, as with the founding
of Cincinnati's dual-language program, was promoted as a means of
attracting immigrant children into the public sector for assimilation.
In 1869, the northern Ohio town on the banks of Lake Erie started
offering German in its common schools, and the Americanization
justification quickly faded away. While pragmatic concerns forced
German to be taught only as a branch of study in many schools, by
the early 1870s nearly half of the schools offering German had a half-
English, half-German curriculum. With the exception of mathematics, all the subjects in the city's select dual-language schools—schools
for which parents had requested German-English instruction—were
taught in both languages. As in Milwaukee, many of the schools were
segregated based on the students' mother tongue, which the supervisor
of the German Department, L. Klemm, saw as "eminently beneficial,
both to teachers and pupils." However, 44 percent of the schools offering German were "mixed," having both native-German and English

speakers. By the 1870s, about a third of the pupils in the city's schools took German, 68 percent of whom were classified as "German." Like other cities dominated by liberal Germans, Cleveland was devoted to innovative education. Object lessons played an important part in the schools' curriculum, as did singing, an important cultural activity for many club Germans. For Supervisor Klemm—a preeminent German educator and writer—the goal of German-English education was cultural advancement. In 1873, he noted that "[t]he better class of Germans of our city see with gratitude that the idea of fusion is not so interpreted as though they were to throw aside all that is peculiar to themselves. They see with pleasure that their good qualities are appreciated. They hold that the fusion of native and German elements and that of the American and German Schools will create a nation, which will be the exponent of the most powerful human culture ever in existence."[13]

In addition to the larger cities, several smaller Midwestern towns also had German-English schools that promoted the aims of German liberals, aims such as mother tongue maintenance, educational innovation, and humanitarian values. New Ulm, Minnesota, and Belleville, Illinois, were two such towns. Disenchanted with American nativism, temperance, and capitalism, a handful of Forty-eighters founded New Ulm in the 1850s as a model community. The town was socialistic, and its schools reflected the liberals' values. Even in such a religious atmosphere as mid-nineteenth-century America, education in the town was fully secular. The New Ulm schools were dual-language institutions, reflecting the liberals' desire for a command of both English and German. Just across the Mississippi from St. Louis, Belleville too was notably progressive. About 80 percent of the town's residents were German, and the *Muttersprache* was introduced in the primary grades in the early 1870s. At first, German in Belleville's public schools was confined to short daily lessons, but the language gradually came to hold a larger place in the curriculum. By the late 1880s, nearly all of the pupils in the town—over 80 percent—learned not only the language, but also the worldview of the German liberals. In addition to the traditional nineteenth-century values of "regularity, punctuality, neatness, [and] obedience," the Belleville schools inculcated their notion of "good citizenship." For the town's school officials, who were mostly German, being a good citizen required "patriotism," but not the sort of patriotism that "fills us with hatred toward other nations or the members of the other political party or sect." Instead, good citizenship—"the positive side of patriotism"—"is the love of everything humane," and this manner of virtuous instruction was what the Belleville schools encouraged.[14]

While German liberals cultivated and shared their values and ideas through a variety of freethinking, Turner, and social clubs, they also established a prominent educational organization, which greatly facilitated their ability to foster a united, broadminded worldview among German Americans. In August 1870, several leading German educators came together in Louisville to form the National German-American Teachers Association (*Nationaler Deutsch-amerikanischer Lehrerbund*). The *Lehrerbund*, which met annually, devoted itself to spreading progressive educational ideas and the German language throughout America's schools. Many of the members of the association came from public school systems in heavily Germanic areas of the country, such as Ohio, Indiana, Michigan, and New Jersey. For instance, Heinrich Fick, the famed supervisor of German in Cincinnati was a leading member. Emmerich and his fellow German teachers in Indianapolis also attended the annual meetings. Other members were affiliated with normal schools, colleges, or semi-public institutions, such as the German American Seminary in Detroit. In the late 1870s, the *Lehrerbund* helped create the National German-American Teachers' Seminary in Milwaukee, a normal school that espoused Forty-eighter ideals. The association noted that "[t]he Germans can offer no better contribution to the people of the United States, besides their industry, than an improved system of education, which, when properly understood and adopted, will have a powerful influence on the intellectual and moral development of the western world." To achieve an "improved" educational system in America, the National German-American Teachers' Association spread information about the latest methods of teaching German as well as other subjects, including geography, mathematics, English, music, drawing, and gymnastics. The *Weltanschauung* the association hoped to promote in the United States included both cultural and intellectual dimensions. Not only was the cultured form of *Deutsch* to be spread and maintained, but a "pupil's power of reasoning" was to be honed as well.[15] Other Germans and other German programs in the United States, however, did not have such lofty and progressive aims.

The "Lower Level" of German Schooling

Although the Forty-eighters hoped to raise the cultural and intellectual level of the German Americans, their desire to create a liberal worldview among the immigrants contained with it a certain amount of snobbery. The "better class of Germans" often looked down upon their less-educated countrymen, noting that they were, culturally, at a "lower level" than themselves. The Forty-eighters, such as Boernstein, noted that these dialect-speaking German immigrants, particularly the *plattdeutsch*-speaking northern Germans, were prone to become "Americanized" and, therefore, "lost forever to Germandom."[16] While the German liberals deliberately cultivated their cultured *Hochdeutsch* and worldview within many public schools in the Midwest, the less-worldly Germans, particularly those in rural ethnic communities, developed public schools that emerged as bilingual institutions organically and without a grand scheme for providing a *Weltanschauung* for the entirety of German America. Moreover, some schools systems adopted German-language instruction in order to do what the Forty-eighters feared most: Americanize the German speakers.

Throughout the Midwest, rural German communities established common schools that became bilingual institutions simply because of the populations they served. In the mid-1880s, the German-American Teachers' Association found that there were over three hundred public German-English schools in the United States, most of which were located in the rural Midwest. The association's estimate was surely conservative because the bilingual nature of many rural schools was often not reported to educational officials. When their dual-language practices were discovered, however, state and county superintendents frequently complained about these bilingual institutions. In the 1880s, for instance, a state superintendent for Missouri lamented that "[i]n a large number of districts of the State the German element preponderates, and as a consequence the schools are mainly taught in the German language." West of St. Louis, in Gasconade County, the German language was used in roughly half of the school districts. The rural Missouri schools sometimes taught in German for a few months and then switched to English for the remainder of the school year; other schools alternated between German and English throughout the school day. Wisconsin too had numerous rural bilingual schools, although, officially, the state only allowed for one-hour daily lessons in languages other than English. The practice of bilingual schools in Wisconsin that "skirted the letter of the law," historian Steven Schlossman has written, "was fairly pervasive, especially from the 1840s to the 1880s." Rural German communities in Illinois also established bilingual schools, which sometimes annoyed state officials. In the late 1870s, for instance, Cook County Superintendent A. G. Lane complained about the "country districts....where Germans have the control of the schools." These rural Germans, Lane noted, maintained their mother tongue in the schools "but neglect the English."[17]

Cook County, of course, was the location of the ever-growing, ever-changing Chicago. "She is always a novelty," Twain wrote of the Windy City, "for she is never the Chicago you saw when you passed through the last time." One thing that was always different about the city was its population; it was nearly as attractive to immigrants as New York. In 1890, German immigrants made up 37 percent of Chicago's 1.1 million residents, making the city home to the second largest population of Germans in the United States. While smaller Midwestern cities had a larger proportion of German inhabitants, Chicago had more Germans than did Milwaukee, Cincinnati, and Cleveland combined. With such a large Teutonic presence, it was not surprising that the city's public schools made the German language part of the curriculum.[18]

Public schooling gradually began to make headway in Chicago during the 1830s and 1840s. In 1865, German was first introduced into the city's public schools, a time when, as historian Paul Rudolph Fessler has noted, the ethnic "vote was up for grabs" due to the political realignments in post-bellum America. Language study in the Windy City began in the first grade, and the number of schools offering the language grew over the subsequent decade. By 1875, however, the study of German was pushed back into the grammar grades. The position of the German language in Chicago's public schools was clearly not secure; it was again allowed at the primary level a decade later, only to be shifted back to the grammar schools in the 1890s. When the Chicago schools merged with several surrounding districts in 1890, some of which were heavily Germanic, the number of schools offering the language increased. But, like New York, Chicago was a multi-ethnic city, and German in the schools, to use historian Steven Schlossman's words, amounted to little more than "token native-language instruction" in order to appease the German element. German was never a serious part of the curriculum, as in other Midwestern cities. When surveying the curriculum of Chicago's schools, educational journals, such as the leading *American Journal of Education*, rarely made reference to the Windy City's German program, but did so with other cities, such as Cincinnati.[19]

Like Chicago, St. Louis's German program differed from those in other Midwestern cities. Public schooling began in the Gateway City in the 1830s, and by the last third of the nineteenth century the city became well known for its educational innovation, particularly with regard to its use of kindergartens. St. Louis, originally a French town, evolved into a German city by the outbreak of the Civil War. In 1860, over half of the city's population was Germanic. The St. Louis Germans were largely clustered on the north and south sides of town; very few native-English speakers, in fact, lived in the southern sections of the city. With such a heavy Germanic presence, the public school officials considered adding German to the curriculum in the 1850s, a decade of intense nativism, but the initiative, unsurprisingly, came to naught. By the 1860s—that pivotal decade for the development of bilingual education—the city's German community petitioned the school board to reconsider German-language instruction. The Germans noted that the language would foster better relations between the Teutonic and Anglo citizens, that "the study of German would naturally assist the study of English," and that "knowledge of the German language" would financially benefit "those who speak it." The school board was swayed by the petition, and in 1864 German was introduced into five schools as an optional branch of study. The German teachers—recruited from Cincinnati's German-English schools—taught short daily lessons in, initially, the German areas of the city. Later, various revisions by the school board made the language available in all sections of the city and in all grades. By 1875, German was an optional program in every public school in St. Louis, with the exception of the schools "for colored children." By the end of the 1870s, the Gateway City's German program was fairly extensive; over twenty thousand students took the language, about a quarter of whom were "Anglo-American pupils."[20]

While St. Louis's German program was booming throughout the 1870s, it was neither intended to cultivate the mother tongue and worldview of the German immigrants as in Cincinnati, nor, as Superintendent William T. Harris noted, a "political concession made merely to conciliate a threatening majority" as in Chicago. Instead, the German language, as Harris made explicit, was a means of "luring" the immigrants into "*our* best institutions" in order to "Americanize" them. If immigrants "establish schools of their own and even achieve a high culture in them, as Germans have done when they have refused to enter *our* public schools," Harris told the National German-American Teachers' Association, "still they may lack training in the spirit of *our* special forms of government," which could be "dangerous politically." Unlike in other Midwestern cities, the Germans were not considered "co-founders" of St. Louis's public institutions; that task was reserved for native-born, English-speaking Americans. Harris noted that "[n]o class of citizens can claim as a right to have any other language than the language of the government taught in the schools at public expense."[21]

Superintendent Harris was not the first to outline the Americanization aims of bilingual instruction in the Gateway City. Assimilation and ending the Germans' private schools had been at the forefront of the St. Louis school officials' minds even before Harris took office. Private schooling was a serious cause for concern because about 80 percent of the German children received that form of education prior to the introduction of German in the public schools. When the school board was considering German-language instruction in the 1850s, Superintendent Ira Divoll noted that such an action would be beneficial because "[i]f the separate and exclusively German schools could be dispensed with, and the German children be induced to attend the public schools more generally than they now do, they would soon become identified in language and in habits of thought and association with the American-born children, and the distinction of nationalities would gradually cease." Although German instruction did not immediately make its way into the public schools, when it did in the mid-1860s, Divoll's mission, later taken up by Harris, was the foundation for foreign-language study. The introduction of German was immensely successful from the perspective of the school officials. By 1880, 80 percent of German children attended the city's public schools. "Large numbers formerly taught by foreign teachers and in private schools," Harris gloated in 1890, "came into public schools and while learning some German have learned much English, to their great future benefit."[22]

The bearded, bespectacled, and philosophic-looking Harris ruled over the St. Louis schools from 1868 to 1880—in 1889 he became the U.S. Commissioner of Education—and made sure his vision of foreign-language instruction was implemented in the Gateway City. Unlike other cities, St. Louis did not create a "German supervisor" position, although there were several "Head Assistants" to supervise the German teachers. Instead, control over the German program was in the hands of the superintendent and the assistant superintendent, and Harris's subordinates fully understood that the goal of the program was to create a "homogeneous people." Louis Soldan, appointed as assistant superintendent in 1870, nearly parroted Harris's words when he, in an overview of the German program in 1871, noted that "[b]y introducing German into the schools, the Board has secured the result, that the future generation grows up as a unit, and imbibes the harmonious national spirit which pervades the public schools of the United States."[23]

Although firm in his conviction that the purpose of German instruction was Americanization, Harris was also a Hegelian and recognized the need for "synthesis." Somewhat of a Germanophile himself, the superintendent of the St. Louis schools did not want German children to completely lose their connection with their high culture. Instead, the Germanic literary and philosophic treasures should be synthesized with the American culture. Harris also wanted to ensure that students' transition to English was not abrupt or absolute because a severed "relation with their family stock would bring calamity" upon both the family and the child. On this matter, Harris was quite prescient and progressive. He hoped to avoid the tumult that often accompanied immigrant children's Americanization by offering some link to their heritage. "[T]he influences derived from communication with the oldest members of one's family," Harris noted, "are very potent in giving tone to the individuality of youth.... This continuity of history is a kind of solid substantial ground for the individual, and from its soil spring up self-respect and aspiration." During Harris's tenure, therefore, the German-language program was partially aimed at lessening the generational conflict that became part of many immigrants' American experience.[24]

The "Flexible" Polyglot Schools

While the Germans—with their large numbers and political clout—were the central ethnic group to have their language taught bilingually in urban public schools, German was not the only non-English language to make its way into the Midwestern common schools prior to the 1880s. In 1876, for instance, F. M. Mason, a township superintendent in Michigan, noted that he was having trouble finding teachers for some of the districts in his neck of the woods. District three in Kawkawlin Township, located in Bay County on the Lower Peninsula, was largely Native American. "Only five of the twenty-seven [pupils]," Mason stated, "could understand a word of the English language, let alone speak it." Therefore, "[i]t was my plan," Mason added, "to put some one in the school who would have patience to teach them by degrees." The township superintendent discovered "an Indian," who was "intelligent, and had the welfare of his people at heart," and immediately certified him as the teacher for the district. Mason did the same thing with a nearby district that was largely French-speaking: he found someone who was capable of teaching bilingually "by degrees."[25] While the French and some Native-American languages became part of the common school curriculum, it was largely the Scandinavians, particularly the Norwegians, who, besides the Germans, used public institutions to maintain their mother tongues in the Midwest. The Norwegian-English schools throughout the upper Midwest followed a pattern of development like the rural German schools; they emerged naturally as bilingual institutions because of the linguistic needs of the communities they served.

By 1880, hundreds of thousands of Scandinavians had emigrated to the United States. Nearly two hundred thousand Norwegians were among those who emigrated, and, along with their American-born children, they constituted a significant foreign-language-speaking element in the upper Midwest were they settled. Over half the Norwegians in America became farmers, largely in Illinois, Wisconsin, Minnesota, and Iowa, as well as farther west on the plains of the Dakota Territory. The Norwegian immigrants were overwhelmingly Lutheran, and like their German counterparts in the Missouri Synod, hoped to give their children a parochial education. Unlike the German Lutherans, however, the Norwegian Americans often lacked the financial resources to begin a parallel system of education in their rural settlements. Instead, the Norwegians largely sent their children to the emerging common schools, and in the rural areas of the upper Midwest, Scandinavians tailored these schools to suit their linguistic, cultural, and religious needs.[26]

The rural common school in the Norwegian settlements, according to O. E. Rolvaag—a prominent Norwegian at St. Olaf College—was "in reality a flexible institution, with all sorts of functions. It served as primary school and grammar school, as language school—in both Norwegian and English—and religious school; in one sense it was a club; in another it was a debating society, where everything between heaven and earth became fit matter for argument; on other occasions it turned into a singing school, a coffee party, or a social centre." These "flexible" polyglot schools that dotted the landscape throughout Illinois, Wisconsin, Minnesota, and Iowa served as typical rural public schools for a few months, especially focusing on reading, writing, and arithmetic, and then, when the regular school year was complete, the schools became vehicles for maintaining the traditions of the Scandinavians. "The primary goal," as one Minnesota teacher noted in 1878, "is to teach the children the Norwegian language, as well as to give them some religious training."[27] The Swedes often used the same arrangement in their rural Midwestern schools. For the Norwegians, teaching the spoken language was not the central aim of the schools since the children had heard and used their dialects from infancy. The real challenge of the bilingual schools was to teach the pupils "to read and write the official Dano-Norwegian norm," which differed significantly from the spoken dialects. A number of textbooks, most of which were published in the upper Midwest, emerged in the last third of the nineteenth century to aid in this process, as well as to convey religious lessons. Although the books surely helped, teaching the official written language was no easy task because, as Einar Haugen has noted, it required the students to "practically learn a new language."[28]

The teachers of the Scandinavian languages in the upper Midwest were often immigrants themselves and poorly trained. In Rolvaag's *Giants in the Earth*, the Norwegian community's school teacher was simply an unmarried man who knew some English. Toward the end of the nineteenth century, more Scandinavian institutions of higher education emerged, and a few trained teachers for both public and private schools. Luther College in Decorah, Iowa, for instance, had a fledgling education department until the 1880s, and the United Norwegian Lutheran Church in Minnesota established a normal school in the 1890s. Yet, keeping school in the "flexible" bilingual institutions was neither easy nor lucrative. While some communities had public school buildings, many educators taught in farms where mothers simultaneously tended to their household duties. Andreas Hjerpeland, who taught in Norwegian schools in the Midwest and West during the 1870s and 1880s, was overjoyed when he found work in a school that had its own building, blackboard, and desks. Typically, Hjerpeland earned twenty dollars a month for teaching and supplemented his income by doing farm labor during the harvest season. He actually made more money working the land, so it was not surprising that he took up farming his 160 acres in North Dakota fulltime in the 1890s, land he acquired with the assistance of the Homestead Act.[29] The patterns of bilingual education found in the Midwest, including the rural, cultural maintenance form used by the Scandinavians, were present farther west as well, although a number of additional languages also made their way into the common schools on the Great Plains and in the West.

4

The Polyglot Boarders Move West

Although the United States was "a mere child," as Theodore Dreiser later described it, by the 1840s the young nation was more than ready to grow—by whatever means necessary. James K. Polk certainly understood the expansionist mood of the country and largely based his 1844 presidential campaign on that sentiment. At the center of America's initial growth spurt was Texas, an independent nation since the mid-1830s and one that desperately wanted to join the Union. Polk's victory demonstrated Americans' passion for expansion, and, therefore, the outgoing president, John Tyler, set in motion the annexation of the Republic of Texas. Americans and their new executive officer, however, had their eyes on more than the Lone Star State. The United States coveted both the disputed territory of Oregon and Mexico's California. Ownership of the lands that would eventually become Washington, Oregon, and Idaho was somewhat unclear; the British and the Americans both claimed them. On his first day as president, Polk unambiguously asserted "the right of the United States to that portion of our territory which lies beyond the Rocky Mountains." "Our title to the country of Oregon," the president insisted, "is 'clear and unquestionable.'"[1] The Oregon question was settled in 1846 through peaceful negotiations with Great Britain, but Mexico certainly would not let go of California without a fight.

In his inaugural address on March 4, 1845, President Polk announced that the borders of America would follow American settlers, even into foreign lands. "The jurisdiction of our laws and the benefits of our republican institutions," the new president stated, ought to reach the settlers "in the distant lands which they have selected for their homes." America's coveting of its Mexican neighbor's lands would ultimately lead to "an evil war," as one observer described the conflict. Although Polk initially attempted to purchase California from Mexico, such an egregious offer was, of course, rejected by the Mexican government. The "evil" war, therefore, was set in motion in order to satisfy America's envious appetite. The precipitating cause of the Mexican-American War was a dispute regarding the border between Texas and Mexico. The United States claimed that the Rio Grande River—and not the Nueces River, which was farther north—made up the border between the nations, and with that justification the two neighboring states went to war. With the signing of the Treaty of Guadalupe Hidalgo in 1848 the conflict was over, and the United States took nearly half of Mexico's lands, lands that would later make up Nevada, California, and Utah, as well as portions of Wyoming, Colorado, New Mexico, and Arizona. In desperate need of money after the war, Mexico agreed to sell part of its remaining northern land to America in the 1850s; the Gadsden Purchase added the southern pieces of New Mexico and Arizona to the United States' holdings.[2]

The West—that conglomeration of lands acquired during a half century of purchase, treaty, annexation, and war—was vast and geographically diverse. Beyond the Mississippi, the Great Plains—a region largely obtained by America from the Louisiana Purchase and the annexation of Texas—was immensely different from the eastern third of the United States. In 1846, an eastern adventurer, Francis Parkman, was in awe when he entered "the great green ocean of the prairies" west of Missouri. The prairie, Parkman noted, was somewhat unsettling because, through it, "[t]he journey was monotonous. One day we rode on for hours without seeing a tree or a bush: before, behind, and on either side stretched the vast expanse, rolling in a succession of graceful swells covered with the unbroken carpet of fresh green grass," grass that was sometimes as tall as travelers' horses. To the north, the plains were fertile and beckoned farmers, while the flora of the lands to the south was well-suited for the grazing of animals, which certainly facilitated the growth of the cattle industry in the region.[3]

West of the Great Plains, of course, were all sorts of geographical wonders, including mountains, valleys, and deserts. In the mountains and along the Pacific coast, a burgeoning fur trade emerged. In the New Mexico Territory, sheep herding became a dominant industry, while the Santa Fe Trail made commerce an important segment of the region's economy. German immigrant Carl Blümner, who worked for one of the trading companies in Santa Fe, was struck by the region's resources. In 1841, he noted that "[t]he land is rich in gold and silver to the highest degree." California was also mineral rich. Richard Henry Dana, a New Englander working on a merchant ship in the mid-1830s, noted that he "never saw so much silver at one time" as when he was in Monterey. In fact, because of a lack of currency in Mexico's northern lands "silver and hides" were used as money, and the sailors referred to these items as "California bank notes." Dana was impressed with California's other natural resources, including its forests, harbors, and fish, but did not care for the "Californians," whom he considered "an idle, thiftless people." But, "[i]n the hands of an enterprising people," he commented, "what a country this might be!" In just over a decade, Dana's "enterprising people," the Americans, would conquer the territory and move to California in great waves, especially after gold was discovered in 1848. "Goldseekers," as Cormac McCarthy has written in his historical novel *Blood Meridian*, were "bleeding westward like some heliotropic plague." Even before the gold rush, however, Parkman—during his adventures on the Great Plains—witnessed a multitude of "emigrants from every part of the country preparing for the journey to Oregon and California." [4]

As the borders of the United States moved west, America's polyglot boarders moved with them. Beginning in the 1830s, Andrew Jackson's Indian-removal policies forced several eastern tribes west of the Mississippi, and beyond the river there were already numerous Native American groups speaking languages other than English. Like Americans in general, the "old" immigrant groups, the Germans and Scandinavians, moved westward, a movement that introduced their languages and cultures to the West. The United States' territorial expansion after the Mexican-American War also made Spanish a significant part of the linguistic diversity in the growing nation. Rather than western emigrants, the Spanish-speaking Mexicans were sometimes northern emigrants, coming to the United States from south of the border. The Chinese too did not move westward, but rather eastward to reach their destination: America's Golden State. With the exception of a few areas, the West was sparely populated prior to the 1880s and, therefore, had fewer public schools than either the Northeast or the Midwest. [5] Nevertheless, each of these foreign-language groups engaged in bilingual education to some extent. While the Midwest was the epicenter of public bilingual schooling in the United States, the West, with its multitude of tongues, also became, educationally, a polyglot region before the Progressive era.

The Old Immigrants

Prior to the outbreak of the Civil War, the American settlement line was still largely in the areas just west of the Mississippi. The Great Plains had not fully been penetrated by whites; it was a region where buffalo outnumbered humans. Before the last third of the nineteenth century, there were several Native American tribes on the plains, as well as wandering mountain men, adventurers, "American backwoodsmen"—clad in "brown homespun," with their "well-beloved" pistols and bowie knives on their sides—and "French Indians," as the long-haired Canadian trappers were colloquially called. By the 1870s, however, more and more settlers moved onto the plains, many of whom were foreign-language-speaking immigrants. Mark Twain once quipped that the driving force for Western civilization's movement was "never the steamboat, never the railroad, never the newspaper, never the Sabbath-school, never the missionary—but always whiskey!" "The missionary," Twain continued, "comes after the whiskey." And there was plenty of whiskey west of the Mississippi. Parkman encountered quite a bit of the liquor on the frontier in the late 1840s, noting that it "circulates more freely...than is altogether safe in a place where every man carries a loaded pistol." According to Twain, after the missionaries have followed the whiskey to the new lands, "next comes the poor immigrant, with ax and hoe and rifle."[6]

Although not really following hard liquor or missionaries, the immigrants from Scandinavia and Germany were seeking land and in their search made their way onto the plains. The move to the Great Plains was greatly facilitated by the Homestead Act of 1862. In 1871, Norwegian immigrant Andreas Hjerpeland explained to a friend "that anyone, be he native or immigrant, has the right, according to the Homestead law, to take a quarter section of land, 160 acres (pronounced 'aker'). You get this by just paying for the papers, about fifteen dollars. When you have lived there for five years, the land is your legal property." By the 1890s, Hjerpeland gave up his occupation as a schoolteacher in order to farm a Homestead tract in North Dakota. The railroad companies also had vast tracts of land that they readily sold to immigrants at a relatively low cost. The Scandinavians, particularly the land-hungry Norwegians, settled in the Dakota Territory, as well as in parts of the Northwest. The ubiquitous Germans—including those from Russia who began to emigrate in the 1870s after the Czar's policy of Russification—settled these areas as well. The immigrants took more than axes, hoes, and rifles to the West; they also took their languages and cultures out on to the frontier, as well as a strong desire to maintain those customs. In order to preserve their culture, the immigrants often settled in rural enclaves. For example, Per Hansa, the protagonist in O. E. Rolvaag's *Giants in the Earth*, noted that he and his countrymen "want only *Norwegians*" at their settlement in the Dakota Territory.[7]

The Dakota Territory, where many Norwegians and Germans settled, was sparsely populated prior to the 1880s and overwhelmingly rural. So thinly inhabited were North and South Dakota that they did not achieve statehood until 1889. In 1880, there were just over five hundred public schools in the Dakotas, but the territory did have a superintendent of public instruction, a position created in 1865. Schooling in the rural ethnic enclaves was often fairly informal. Held in a community member's house or perhaps in a soddie (because of a lack of timber on the plains), the school convened during the winter months after the crops had been harvested, and it gave the students the rudiments of learning. As the public school system developed in the Dakotas, bilingual instruction emerged naturally, as in the rural areas of the Northeast and Midwest. Although both the Scandinavians and the Germans of the northern plains sought to maintain their languages, the Russian Germans, in general, clung even tighter to their linguistic traditions. These Germanic immigrants had become accustomed to shielding their heritage from the onslaughts of Russian culture in the Old World.[8]

The bilingual schooling of the Scandinavians, the Germans, and, to some extent, the French was somewhat unsettling to the superintendents of education in the Dakota Territory. Although English was supposed to be the language of instruction in the territory, one school official complained in the 1880s that "nothing could be done about it, as the foreign element was so strong that they not only controlled the schools but the election of the county superintendent also, and a strong public sentiment was created in support of the schools taught in a foreign language." Even into the 1890s, the state-level curriculum, which emphasized English, was difficult to implement because of the "large foreign population." In North Dakota, a Logan County superintendent noted that "[t]he course of study for this state is followed as closely as possible," but added that "it is very difficult" because the "county is settled almost entirely by German Russians." Other county superintendents complained about the difficulty of putting into practice the English course of study and lamented the control of schools by German-, Scandinavian-, and French-speakers.

Some superintendents, however, recognized the necessity of bilingual education in the Dakotas. V. H. Stickney, superintendent for Stark County, North Dakota, noted that "[i]n some of the schools...out of the eight or ten pupils that made up the list of attendance, not a single one...could converse intelligently in English. In such schools I have endeavored to have the school boards...secure the services of teachers who could speak both German and English." Although public bilingual schooling on the northern plains emerged organically because of the foreign-language-speaking populations in the Dakotas, it came to subvert the intentions of state officials as the public schools became more organized. As one North Dakota superintendent of public instruction put it, "This is our beloved America—no[t] provinces of [the] German, Scandinavian, or French" peoples; given the linguistic diversity on the Great Plains, the superintendent was perhaps somewhat mistaken.[9]

Southward on the plains, foreign-language speakers—"Swedes" as "all light-haired men who spoke with a heavy tongue" were colloquially called—entered into areas that were, at times, more considerate of their desires for linguistic and cultural maintenance. While the Scandinavians tended to settle in the Dakotas and farther west in Oregon, the Germans went south as well. German immigrants formed significant pockets of foreign-language speakers in Nebraska, Kansas, and Texas. Like the Dakotas, these states were, for the most part, sparsely populated prior to the 1880s, and the rural pattern of bilingual schooling predominated. In the states south of the Dakotas, however, bilingual education in the public schools was sometimes guaranteed by law. Kansas's school law, for instance, allowed for the German language in the state's common schools from 1867 to 1874. By the early part of the twentieth century, Nebraska followed suit and, for five years, permitted any foreign language to be taught in its schools. Texas—with its concentrations of Germans in Galveston, Houston, and San Antonio, as well as in a host of rural areas—largely skirted the language issue in its school law prior to the 1870s. During the Reconstruction era, however, the Lone Star State allowed for German, French, and Spanish in the Texan schools, although the law, which was amended regularly, stipulated that no more than two hours daily should be devoted to the languages. In 1871, Texas appointed its first superintendent of public instruction to implement the new school law. Superintendent Jacob C. De Gress, an immigrant from Germany, interpreted the bilingual provision as a policy geared toward enticing the state's foreign-language speakers into the public schools for assimilation. Because localism prevailed, not all German communities followed the intent of the law. In some rural schools, half of the curriculum was taught in German, and in towns such as New Braunfels teachers used German as a means of promoting a more progressive pedagogy, not as an instrument of Americanization.[10] In Texas alone, therefore, multiple patterns and aims of bilingual education simultaneously existed with regard to German-English instruction, but the most significant aspect of bilingual schooling in the Lone Star State centered not on the "old" immigrants, but on the colonized Spanish-speaking populations.

Los Mexicanos

Spain's Mexican colony had been a highly stratified society, one in which the European-born rulers lorded power over the American-born creoles, the mixed-blood mestizos, and the indigenous populations. Under the leadership of Agustín de Iturbide, the creoles—jealous of the Spaniards' power but not willing to change the hierarchical structure of society—generated a conservative revolution and gained independence in 1821. As an independent nation, however, Mexico was not fully stable. Liberals and conservatives battled for power, setting in motion something akin to the "eternal war" Nobel Prize–winner Gabriel García Márquez described in his *One Hundred Years of Solitude*. Mexico's northern frontier—the areas later conquered by the United States—was largely untouched by the wars. The political and economic situation in *el Norte* remained unchanged for the most part, and the reforms of the new governments often did not reach the northern rim of Mexico. This was especially true in the relatively isolated lands that would become New Mexico. "Independence from Spain, and the Mexican regime that followed," education scholar George I. Sanchez has noted, "had little effect upon the life of the New Mexican." Therefore, "the New Mexican has been forced to live in a world of his own making," a world that was rooted in the colonial era and one in which "[t]he rhythm of centuries had set a persistent tempo to...life." As in Márquez's mythical novel, it seemed as though "time was not passing"; "it was [merely] turning in a circle" over the years.[11]

Like the earlier Spanish colonies, the northern sections of the Republic of Mexico were highly stratified. Political and, especially, economic power rested in the hands of a few elite families in the regions that would eventually become the American Southwest. The

Spanish Dons in California and the *ricos* in New Mexico controlled vast tracts of land and relegated the lower orders, many of whom were mestizos and Indians, to an almost feudal form of serfdom and patronage. A member of the elite Vallejo family in California remembered the time "before the American conquest" with a certain amount of nostalgia. A "quiet and happy domestic life," noted Guadalupe Vallejo in 1890, was the norm "when only a few Spanish families" dominated California; the "Indians," Vallejo asserted, happily attended the needs of the elites, even "singing hymns" as they worked. After the conquest, however, those formerly in power faced a new and daunting world. "Naturally," Willa Cather fictionally wrote of an influential priest in Mexico's northern lands, "he hated the Americans. The American occupation meant the end of men like himself. He was a man of the old order,...and his day was over."[12]

The economic and political stratification of society in Mexico's northern territories was closely connected to race. During his visit to California in the 1830s, Dana noted "each person's caste is decided by the quality of the blood." The Mexicans' "complexions," therefore, determined "their rank" in society. "Those who are of pure Spanish blood," Dana continued, "having never intermarried with the aborigines,....form the aristocracy; intermarrying, and keeping up an exclusive system in every respect." This Mexican racial hierarchy was not as rigid as that in the United States, however. The presence of white "blood" among African Americans did not elevate their status. Frederick Douglass, for instance, remained in bondage until he eventually fled to the North, even though his "father was a white man." In Mexico, by contrast, "the least drop of Spanish blood," Dana reported, "is sufficient to raise them [Indians] from the rank of slaves...and to call themselves Españolos, and to hold property."[13]

The American obsession with racial purity did not bode well for the Mexicans in the Southwest after they were conquered by the United States. When General Stephen Watts Kearny and his soldiers—en route to Santa Fe—took control of Las Vegas in 1846, he suggested that Americans harbored no ill feelings toward the inhabitants, promising that he and the soldiers were "friends—not...enemies." "[T]he laws of the United States," Kearny stated, will "extend over" the conquered territories, and "those who remain peaceably at home, attending to their crops and their herds, shall be protected by me, in their property, their persons, and their religion." The Treaty of Guadalupe Hidalgo, which ended the conflict in 1848, also hinted that the colonized Mexicans would hold the "title and rights...of citizens of the United States." The American tradition of racism toward non-Westerners, however, ensured that the Spanish-speaking populations would, for the most part, become second-class citizens. In 1848, for instance, Senator John C. Calhoun opposed the American expansion into the Southwest because "we have never dreamt of incorporating into our Union any but the Caucasian race—the free white race." "We have conquered many of the neighboring tribes of Indians," the senator continued, "but we never thought of…incorporating them into our Union[;]….more than half of the Mexicans are Indians, and the other is composed chiefly of mixed tribes. I protest against such a union as that! Ours, sir, is the Government of a white race." Writing in the 1930s, historian Walter Prescott Webb also concluded that many northern Mexicans were an inferior people because, in the colonial era, those "in the Spanish service came largely from pueblo or sedentary Indian stock, whose blood, when compared with that of the Plains Indians, was as ditch water." Parkman, living among the Native Americans on the plains some eighty years before Webb's analysis, came to a similar hierarchical view, noting that "[t]he human race in this part of the world is separated into three divisions, arranged in the order of their merits: white men, Indians, and Mexicans."[14]

Being a supposed inferior race, *los mexicanos* were subjected to all sorts of negative stereotypes, and the stereotypes helped justify not only the conquest of Mexico and colonialism, but also the taking of the best lands in the Southwest by the Americans. One of the central perceptions of the Americans was that Mexicans had been struck by "laziness," what Dana called "California fever," a phrase that would greatly change in meaning after the discovery of gold. In the 1840s, Parkman characterized the Mexicans as a lazy "race," as did Rufus B. Sage, a New Englander visiting New Mexico. Sage also made note of the "ignorance" and poverty of the New Mexicans, conditions that, in the popular mind, were intimately linked. As a character in a novel by Orhan Pamuk—the celebrated Turkish writer—poignantly stated, "Mankind's greatest error…is…to confuse poverty with stupidity." "[W]hen an entire nation is poor," the Kurdish character noted, "the rest of the world assumes that all its people must be brainless, lazy, dirty, clumsy fools…. [T]heir culture, their customs, their practices" then become "a joke." The hierarchical structure of the society was often cited as evidence of the Mexicans' laziness. "Among the Spaniards," Dana observed, "there is no working class; (the Indians being slaves and doing all the hard work…)." Interestingly, José María Sánchez, visiting Texas from Mexico City in the late 1820s, also found Mexico's northern inhabitants—compared to those in the southern urban centers—to be unenlightened and lazy, noting that "the worst punishment that can be inflicted upon them is work." While some of the *tejanos* shunned hard labor, none was lazier than the American settlers streaming into Texas. Almost mirroring Dana's description of the *californios*, Sánchez noted in 1828 that the Americans "are in general...lazy people of vicious character. Some of them cultivate their small farms by planting corn; but this task they usually entrust to their negro slaves, whom they treat with considerable harshness."[15]

Some Mexican Americans, particularly the elites in the Southwest, attempted to undermine the Anglos' negative stereotypes by asserting their European roots. Prior to the turn of the twentieth century, for instance, members of the Spanish-speaking intelligentsia in New Mexico began promoting a *hispanidad* movement, which not only generated a pride in their Spanish heritage but also, for some, told the Anglo world that they too were "white" and, therefore, were being marginalized and stereotyped unjustly—unjustly according to the racist sentiments of the nineteenth century. As part of this movement toward claiming an "invented" racial identity, as historian John M. Nieto-Phillips has called it, many colonized Americans in the Southwest referred to themselves as "Spanish Americans." In California, Guadalupe Vallejo made sure to identify family members as "Spanish Californians," which, according to Vallejo, were distinct from the "Mexican...and Indian population[s]." In fact, this "fantasy heritage" was largely incorrect. The "Spanish-American" view that the elites were direct descendants of European colonizers and were, therefore, racially "pure"—as well as the reports from travelers like Dana that supported that myth—ignored the widespread intermingling of Indian and Spanish "blood" at all levels of society.[16]

Although the negative stereotypes made them appear monolithic, *los mexicanos* in the American Southwest were a diverse people. Not only was the Spanish-speaking population divided by class—including a developing middle class, which accounts of the Southwest often missed—but by geography as well. The *tejanos*, *nuevomexicanos*, and *californios* had largely lived in isolation from one another, and distinct commercial endeavors and modes of life characterized each sub-region. In 1828, Sánchez was struck by how different the *tejanos* were from the Mexicans to the south, a difference that was made all the more salient because of the arrival of the Anglos. The Mexicans in Texas adopted some of "their customs and habits" from the Americans, and, Sánchez noted, "one may say truly that they are not Mexicans except by birth." In addition to the Spanish speakers forcibly incorporated into the United States after the Mexican-American War, tens of thousands of Mexicans—the exact figures are unknown because, then as now, the borders were quite porous—migrated northward during the four decades after the war and brought even more diversity to the Mexican-American population in the Southwest. These newcomers, often poor laborers, were sometimes viewed by the Americans as inferior to the "Spanish Americans" who dominated the region prior to the American conquest. For example, W. W. Mills, an Anglo resident of El Paso during the second half of the nineteenth century, noted that the older Mexicans "were of a much better class than those who came in later with the advent of the railroads, to sell their labor—and their votes." Although not especially pleased with the arrival of the Mexican newcomers, Mills, however, recognized that they were being wrongfully exploited by whites. "[V]otes cannot be sold unless there be purchasers," Mills noted, "and...the purchasers have ever been of my own race."[17]

Although the Mexican-American population was marked by diversity, what the Chicanos in the American Southwest shared was a mother tongue and a devotion to Catholicism, both of which were clung to tightly. Before the Mexican-American War, Anglo settlers were expected to learn *español* and be baptized by the Church, conditions to which Stephen F. Austin and his companions had to agree before being allowed to move into Texas. In Santa Fe, Blümner reported in 1841 that only Catholicism "may be practiced within the borders of the *Republik*." According to Dana, the situation was the same in California, noting that a Protestant sailor "must leave his conscience at Cape Horn." Although Dana was not particularly fond of the *californios*, he was enamored by their language, which he soon learned after arriving in Monterey. He noted, "It was a pleasure simply to listen to the sound of the language," and because of the Mexicans' poverty "they sometimes appeared to me to be a people on whom a curse had fallen,...stripped...of everything but...their [beautiful] voices." Coming from Mexico's interior, Sánchez was not impressed with the language of his unenlightened northern countrymen, noting "they even speak Spanish with marked incorrectness."[18]

Because language and religion were so central to Mexicans' heritage, the Treaty of Guadalupe Hidalgo guaranteed both in theory, although not always in practice. Article IX "secured...the free exercise of their [Mexican Americans'] religion without restriction," and the treaty's assertion that the culture of the new citizens would be protected was interpreted, particularly by those in New Mexico, as an assurance that the Spanish language would be cultivated in the public schools. Because of their devotion to Catholicism, however, many Mexican Americans maintained their language in parochial schools. In New Mexico, for instance, the system of Catholic schools prior to the 1890s was something of a bulwark against the full development of public schools in the territory. In addition to the religious schools, in which the language of instruction was frequently Spanish, Mexican-American cultural leaders often established private schools, some of which were bilingual. Not unlike the activities of the "club Germans," middle-class Spanish speakers in the Southwest founded mutual aid societies during the last quarter of the nineteenth century, and these *mutualistas* not only provided support for Mexican Americans, but also sought to promote Mexican culture through independent schools, Spanish-language newspapers, and libraries. In the 1880s, Thaddeus M. Rhodes, a judge in Hidalgo County, Texas, reported that "in many of the communities private Spanish schools are established, which materially interfere with the public schools."[19]

In the former Mexican lands, public bilingual schooling for Spanish-speaking children was most pronounced in Texas and New Mexico and was nearly ignored in the other southwestern states and territories. Because of its sparse population, common schooling did not emerge in Arizona until the 1870s. By 1880, the state could claim only about one hundred public schools. From its inception, public schooling in Arizona was largely an English-only endeavor. In 1890, the teaching force in the state was almost exclusively composed of Anglos, Pedro Michelena of Graham County and Teofilo Aros of Pima County being among the few exceptions. Although German was permitted in Colorado's schools since 1867, Spanish did not achieve the same status until a decade later. While state school laws recognized German and Spanish, they were only allowed to be taught as a subject, and school officials made clear that English was to be the language of the schools, even in the heavily Spanish-speaking areas of the state. In 1888, Huerfano County Superintendent Fred Pischel lamented that the "Mexican population" was "not making as rapid an advancement in the English language as the importance of the case would lead us to expect," largely because the students, the superintendent speculated, spoke Spanish at home. Pischel wondered, "Where lies the remedy?" In California, the western migration, intensified by the rush for gold, made the state largely Anglo in composition. A number of schools in the Golden State added non-English languages to their curricula, but the languages tended to be those of the older American immigrant groups, especially German. Like Colorado, California largely emphasized English in its public schools. As a committee on the California curriculum noted, "The primary schools must not fail to secure to their pupils as thorough a training in the English language as time and the ability of little children will permit."[20]

In Texas, by contrast, public bilingual schooling flourished prior to the 1880s. As with German, Spanish received state-level support during the Reconstruction era. Although guaranteed by law, bilingual instruction in the Lone Star State was interpreted as a tool to attract foreign-language-speaking students into the public schools in order to assimilate them to the American language and culture. After Reconstruction, however, when the Democrats once again gained control of politics, they moved toward localism with regard to education. Therefore, local officials, particularly county judges, determined the language of instruction for the public schools. With localism reigning supreme, common schools reflected their surrounding communities, and Spanish-English schooling became prevalent in the southern and western sections of Texas, areas that had large numbers of *tejanos* and *mexicanos*. For instance, El Paso, with its large Mexican-American population, was a bilingual town, and it schools came to replicate the practice. Writing about El Paso during the second half of the nineteenth century, Mills noted that "Spanish was the language of the country, but many of our Mexican friends spoke English well, and often conversations, and even sentences, were amusingly and expressively made up of words or phrases of both languages." Reflecting the larger bilingual community, El Paso's public schools naturally became dual-language institutions. Moreover, El Paso's closeness to New Mexico, with its strong promotion of Spanish culture, clearly influenced the border city's commitment to bilingual instruction.[21]

Because localism was the norm in Texas during the post-Reconstruction years, the nature of bilingual education depended on the whims of local school officials. Farther down the Rio Grande River from El Paso, where numerous Mexican Americans settled, bilingual instruction was fairly common. In Val Verde County's schools, for instance, the language of instruction was in both Spanish and English until W. K. Jones, a county judge, implemented an English-only policy in the 1890s. Southward, Zapata County also offered bilingual schooling to its largely Mexican-American population. At the southern tip of Texas, Hidalgo and Cameron Counties attempted to transition their Spanish-speaking students to English. E. H. Goodrich, the county superintendent for Cameron County, noted in the 1880s that the Mexican-American children's "only knowledge of the English language is obtained in the public schools," which was seen as an "obstacle" to the aims of the school leaders: an intimate familiarity with the American language. Up the coast in Nueces County the aims of bilingual education were much the same as in Hidalgo and Cameron Counties. County Judge Joseph Fitz Simmons stated in the late 1880s that "[t]hree-fourths of the scholastic population west of the city of Corpus Christi being of Texas-Mexican origin...necessitates the employment of teachers having a knowledge of both the English and Spanish languages." Although bilingual education was utilized, Judge Simmons continued, the aim of the dual-language instruction ultimately was "Americanizing the rising generation."[22]

In the neighboring territory of New Mexico, the Spanish language was maintained to a high degree, largely because the *nuevomexicanos* were able to resist assimilation. Unlike California and Texas, New Mexico was not a major destination for Anglo settlers, which enabled Spanish speakers in the territory to hold on to their customs more readily. New Mexico was also the center of the *hispanidad* movement, a movement devoted to the cultivation of the "Spanish-American" culture. Although the notion of a "Spanish-American" heritage—distinct from and, therefore, superior to Mexican-American culture—was largely a myth, the *hispanidad* movement generated a great deal of pride in the Spanish language throughout New Mexico. The Spanish-language newspaper *El Labrador*, for instance, told its readers at the turn of the twentieth century that "[t]he rights of our citizens in that which appertains to their language and customs are protected by the laws and the treaties and none has the right of abolishing them or limiting them." Furthermore, *El Labrador* stated, "the native people of New Mexico do not have reason for being ashamed of their language. Honor and self-respect should impel them to conserve it."[23]

Schooling, of course, was one of the avenues of conserving the language, but public education was largely neglected, thus allowing private schools to take care of most of the formal education in New Mexico prior to the late 1880s. In 1870, the territory had only five public schools; a decade later there were over 150, still a relatively modest number. By the early 1890s, public education became a priority, which, territorial leaders understood, was needed in order to achieve statehood. New Mexico, therefore, created the position of superintendent of public instruction, and Amando Chaves filled that office from 1891 to 1897. Given New Mexico's devotion to the preservation of the Spanish language, many of the early schools were bilingual. In 1889, for instance, 28 percent of the territory's public schools were Spanish-English institutions, while 42 percent used English exclusively and 30 percent had a Spanish-only curriculum. In 1891, a new school law codified the utilization of either Spanish or English (or both) in the school system, and Superintendent Chaves actively sought out bilingual teachers for the territory's elementary schools. Although *public* education played a restricted role in New Mexico prior to the 1880s, the territory's dedication to the cultivation of Spanish and to bilingual instruction provided a bulwark against the English-only policies that became more pronounced throughout the Progressive era.[24]

The Native Americans

During his famous visit to the United States in the 1830s, Alexis de Tocqueville reported that Americans continuously move into Indian lands because of the lure of potential financial gain. "[T]he desire for prosperity," Tocqueville stated, "has become a restless and burning passion which increases each time it is satisfied....Emigration for...[the Americans] began as a need and has, today, become a sort of gamble which rouses emotions which they like as much as the profit." By the 1820s, the Americans' insatiable desire for profit compelled the federal government to force many of the Native Americans of the Southeast farther westward in order to free up land for white settlement. Andrew Jackson, whom Tocqueville described as "a man of a violent disposition and mediocre ability," supported the Indian Removal Act of 1830, which authorized the taking of Native American holdings in the Southeast in exchange for lands west of the Mississippi. In 1831, Tocqueville witnessed the "fearful evils" of the "forced migrations" of the Choctaws across the Mississippi in "the depths of winter." Like the Choctaws, the Cherokees, after losing battles for their ancestral lands in the Supreme Court, were also forced west, and the march on the "Trail of Tears" killed almost a fourth of the tribe. By 1870, of the nearly 400,000 American Indians, only about 30,000 inhabited the regions east of the Mississippi. Forcibly, therefore, Native Americans had become residents of the West.[25]

Of all the ethnic groups in the United States perhaps none was more diverse that the American Indians. Largely concentrated in the West by mid-century, the Native Americans were differentiated by tribe, custom, and geography. The "Five Civilized" tribes of the Southeast— the Cherokee, Chickasaw, Choctaw, Creek, and Seminole—were forced into "Indian Territory" in Oklahoma—"the Nations," as the Cherokee referred to the region. The Arapaho, Cheyenne, Kiowa, Pawnee, and Sioux historically inhabited the Great Plains, while the Navajo, Pueblo, and Hopi were numerous in the Southwest. Some tribes relied on big game for sustenance, while others cultivated the land or raised livestock. Languages, comprising over twenty distinct linguistic "families," too contributed to the diversity among American Indians. The Cherokee, Navajo, Lakota, and Dakota—the latter two being Siouan—languages were used by large numbers of people and, eventually, were put into written form. Not every tribe had its own unique language, however. The Omaha, Kansa, Quapaw, Osage, and Ponca, for instance, all spoke the Dhegiha form of the Siouan tongue. Some large groups spoke a multitude of languages, such as the Pueblo Indians who spoke Keres, Tewa, Tiwa, Towa, Zuni, and, frequently, Spanish. The Plains Indians, who utilized a variety of different dialects, often shared a common sign language for inter-tribal communication. In addition to the differences that existed between groups, individual Indians' standing within the tribes generated more layers of diversity among America's native inhabitants; some nations, such as the Cherokee, had intricate hierarchical structures based on racial status and wealth.[26]

For many nineteenth-century whites in the United States, much of this diversity was lost; Indians, regardless of tribe or position, were "savages." While recognizing the distinctions among the various tribes on the Great Pains, Parkman concluded that American Indians were, in general, "savages," although savagery was a matter of degree. In 1879, the Commissioner of Indian Affairs agreed that "all the Indians...have been savages," but noted that some were beginning to shed their uncivilized ways. For Parkman, Native Americans were not only savage, "all Indians" shared the tendency to lie, brag, and believe in "absurd" superstitions. Given these sweeping negative generalizations that dehumanized Native Americans, "Indian-hating" became relatively common in the West. Parkman, for instance, encountered a Kentuckian whose main objective in his western travels "was to kill an Indian"; the young man eventually murdered a Pawnee—like Cather noted of the "hunt" for Navajos—as "a form of sport."[27]

Savagery, American political leaders agreed, had to be contained and, eventually, overcome. The eastern tribes had largely been subdued through their internment in Indian Territory, but the Plains Indians and the western tribes were still untamed. In the 1850s, because of the growing danger to the western travelers, many of the Native Americans on the Great Plains were restricted to specific areas of the West, setting the stage for the reservation system that was to contain savagery. By the 1870s, forcing American Indians in the West onto reservations was a large-scale federal policy. Enormous reservations were established not only in Oklahoma, but also in Montana, the Dakotas, Nebraska, Colorado, New Mexico, and Nevada; the Native Americans had become dependent "wards" of the federal government. Although confined to reservations, savagery still needed to be defeated. As the Commissioner of Education made clear in 1870, "[e]ither citizenship or extinction seems to be the Indian's destiny."[28] For many years, extinction was more than just tough talk; it was a live option for many of America's political leaders, and, as a reflection of this attitude, Indian policy initially was the responsibility of the Department of War. With "[s]ome of the Apache bands," reports from the Southwest indicated, "war will continue, in all probability, until they are exterminated." Numerous battles were fought against the Indians' supposed savagery, but in the decades following the Civil War the genocidal tendencies of the federal government seemed to wane somewhat, particularly after such cruelty came under attack by post-bellum reformers, Indian advocates, and concerned political leaders. By the late 1870s, the Department of the Interior, in which Indian Affairs was housed after 1849, came to reflect the trend toward more humane treatment of Native Americans. In 1879, Secretary of the Interior Carl Schurz—perhaps the best-known of the liberal German Forty-eighters—stated that "it is our solemn duty to leave nothing untried to prepare a better fate than extermination...for the original occupants" of the United States.[29]

Although bringing "civilization" to the "savages" appeared to be the only humane alternative to the "Indian problem," the path to such civilization was neither an easy nor a rapid one. "Civilizing" the Native Americans, in fact, involved a multigenerational and multifaceted approach. Primarily, American Indians needed to value private property by becoming independent cultivators of the land. During Rutherford B. Hayes' presidency, the first goal of Secretary Schurz was "[t]o set the Indians to work as agriculturists or herders, thus...break[ing] up their habits of savage life and...[making] them self-supporting." By the 1880s, the Dawes Act, which divided reservations into individual plots, attempted to make Indians the sort of yeoman farmers of whom Thomas Jefferson would be proud. In addition to land allotment, the rising generation needed an education that would strip it of its savage ways. Education, Secretary Schurz noted in 1879, was the second most important aim of the federal government's "Indian policy," because it inculcated "civilized ideas, wants, and aspirations." Critical of these more "humane" policies, G. Stanley Hall, the famed psychologist and president of Clark University, stated, "Instead of trying to make a good Indian, we try to make a wretched third-class white man."[30]

Between the 1840s and 1880s, there were a number of publicly financed schools that sought to cultivate "civilization" among the American Indians. While local control of education was the norm prior to the Progressive era, such an important task as civilizing the Native Americans could not be left to the whims of any particular local community (although the state of New York had its own Indian schools); the federal government, therefore, shouldered "the white man's burden" of taming the "savages" through public schooling. As wards of the United States, the American Indians largely received an education that was both federally funded and controlled. The government supported both day schools and boarding schools on the reservations. By the 1880s, off-reservation boarding schools came to be seen as the most effective means of civilizing Native Americans, especially after the successes of Richard Henry Pratt's Carlisle Indian School in Pennsylvania. The federal government also partially financed several missionary schools, although some of these sectarian schools received only private funds. Moreover, the Cherokees, Chickasaws, Choctaws, Creeks, and Seminoles were allowed to maintain their own public schools in Indian Territory, presumably because they were considered "civilized" tribes. In 1884, John B. Riley, the superintendent of Indian schools, reported that there were 95 boarding schools and 80 day schools funded by the U.S. government, and the tribes in Indian Territory operated 17 boarding schools and about 160 day schools.[31]

For the U.S. government, "civilization" was intimately linked to the English language. In the 1870s, Secretary Schurz noted that "Indian...children [must be] instructed in the ways and arts of civilized life, and especially in the English language." By the 1880s, when English-only instruction became a federal goal for Native Americans, the superintendent of Indian schools claimed that the Indian's "inability to speak another language than his own renders his companionship with civilized man impossible." According to the government, English was the key to ending savagery among the American Indians. "Until their barbarous dialects have given way to civilized language," the superintendent argued, "the duties of citizenship…will be beyond their comprehension." "In fact," it was noted, "the Indian who speaks the English language is a savage no longer."[32] Even the public schools operated by the five "civilized" tribes had become, at least officially, English-only institutions. In 1884, Superintendent Riley reported that the tribes' "common-school system[s], in which the books used, as well as the instruction, are required to be in the English language exclusively." This English-only aspect of Indian education was disconcerting to students with no knowledge of America's predominant language, especially when it was harshly implemented. For example, when Francis La Flesche and his Omaha classmates began school in the 1860s, they "encountered a rule that prohibited the use of our own language, which…was rigidly enforced with a hickory rod, so that the new-comer…was obliged to go about like a little dummy until he had learned to express himself in English." Not only were students required to speak English, they were also given English names because, as La Flesche reported, "their Indian names were difficult for the teachers to pronounce." Luther Standing Bear, a Sioux at the Carlisle Indian School, had to pick a new name from a list on the blackboard. "None of the names," Standing Bear noted, "were read or explained to us, so of course we did not know the sound or meaning of any of them."[33]

The link between English and civilization—which began to emerge in the 1860s, but became official policy in the 1880s—hampered bilingual instruction among the Native Americans, but it was not the only obstacle. Native languages were also hard for the whites who operated Indian schools to learn. Parkman, for instance, noted that the "language of the Arapahoes [sic] is so difficult, and its pronunciation so harsh and guttural, that no white man, it is said, has ever been able to master it." Several missionaries, nevertheless, did master native languages, and made use of bilingual instruction in their schools. John Homer Seger, who taught at the Cheyenne-Arapaho Agency in Oklahoma for decades beginning in the 1870s, even learned the difficult Arapaho tongue and spoke it to win over the students and parents to the value of education. Stephen R. Riggs, a missionary among the Sioux, learned the Dakota tongue and published a grammar of the language in the 1850s. Riggs' schools, including the Santee Sioux Normal School, were bilingual; the schools utilized books translated into Dakota, which Riggs argued facilitated students' transition to English. Occasionally, Catholic, Baptist, Methodist, and Quaker missionaries also used bilingual instruction in their schools. Even in government schools, bilingual education, although discouraged, was a regular aspect of the curriculum. The superintendent of Indian schools reported in 1887 that the students in some schools were allowed to make use of their native languages "except when reciting"; the superintendent, however, argued that bilingual instruction interfered with students' English acquisition, noting that "although they had been in school for several years, they could not speak so as to be understood." Although some missionaries and Indian school leaders used bilingual instruction, it was not intended to preserve vernacular tongues. Instead, like the school systems that used German-English instruction to Americanize the immigrants, bilingual education in missionary schools aimed at little more than quickly converting Native Americans to Christianity. By the late-nineteenth century, however, government-sponsored missionary schools were on the wane—largely because of conflicts between Catholics and Protestants over funding—and the federal government thus monopolized the education of American Indians. The ending of missionary schools was even supported by the pan-Protestant organization Friends of the Indian in order to quell the Catholic presence in Indian education.[34]

While bilingual instruction was occasionally a school-wide endeavor, it occurred at a more individual and informal level in several Indian schools as well. At the Carlisle Indian School, which was an English-only institution, an interpreter initially was employed in order to help the newcomers—with no knowledge of English—understand the rules of the boarding school. After the students understood the rudiments of the American language, however, Pratt quickly let the interpreter go because he wanted Carlisle "to become an English-speaking school as quickly as possible" and assumed that Native Americans "could learn faster if forced to use" English "all the time." Standing Bear, who had attended Carlisle, began his own teaching career at an Indian agency school on the plains in the 1880s. Although English was the language of instruction at the school, when "children read a line of English, and...did not understand all they had read," Standing Bear "would explain it to them in Sioux." In the Cherokee, Choctaw, and Creek schools, in which English was the language of instruction, individual teachers, many of whom were Indians, likely aided students in their native tongues when they failed to understand, as Standing Bear did with his pupils. In addition to interpreters and teachers, students themselves often became informal bilingual instructors. At the mission school La Flesche attended in Nebraska during the 1860s, older students often spoke to nervous newcomers "in Indian" in order to console them and to "help them along" in their English understanding.[35] By helping others to survive in a frightening atmosphere, this informal bilingual instruction among Indian teachers and students demonstrated both their ethnic solidarity and their willingness to resist the whites' educational imposition.

The Chinese

Like Native Americans, the Chinese immigrants who began to arrive on America's western shores during the mid-nineteenth century were perceived as racially inferior to whites and, therefore, morally degenerate. In 1870, H. N. Day, an opponent of anti-Chinese sentiment, reported to the U.S. Commissioner of Education that the "paganism and superstition" of the Chinese need not be feared because there was "little danger of its infecting our native population; little danger of its spreading at all among us." Others, however, were not so optimistic about Chinese culture. Eight years later, a report submitted to the U.S. House of Representatives stated that the Chinese immigrants' "personal habits, peculiar institutions, and lax morals render them undesirable members of society." In fact, the "coolies," according to House's Committee on Education and Labor, "are impregnable against all influences, and remain a quiet, united class, distinct from us in color, in size, in features, in dress, in language, in customs, in habits, and in social peculiarities." "Indeed," the committee noted, "the only purpose in society for which they are available is to perform manual labor."[36] While the American Indians faced the ultimatum of either civilization or extinction, the Chinese confronted—as H. R. Rice, a Presbyterian minister in Sacramento, explained—the choice of being "educated or excluded." Although in the late 1870s Rice concluded that it was not "possible to exclude them," the Chinese Exclusion Act of 1882 attempted to do what Rice thought unachievable. The Chinese living in the United States, like the Native Americans, had very little control over their education in the public schools because, as Rice noted, it was "the duty of the Government" to teach the Chinese the "American customs...in order to civilize them."[37]

Although a sprinkling of Chinese immigrants came to the United States before mid-century, the flood gates of Asian emigration opened in the 1850s, a decade in which about forty thousand Chinese arrived in the United States. By 1870, over a hundred thousand immigrants made the voyage to America, a voyage that, because of increased competition, sometimes cost as little as twelve dollars. The Opium Wars and a failing economy in China prompted this large-scale move to America's recent conquests in the West, areas that were experiencing an economic boom. Unlike the "old" European immigrants, many of the Chinese newcomers did not intend on settling in the United States permanently; most of the Asian immigrants were men seeking economic betterment in order to eventually return to their families in their homeland. These Chinese men in America were a diverse lot. Although the stereotype was that they were all "coolies"—forced laborers—many of the immigrants were in fact free and came of their own volition. There were members of the "higher and titled classes" among the newcomers as well. While the majority of the Chinese immigrants were laboring men, most were literate according to Day. Initially centered in California, the Chinese men worked on the railroads, on farms, and in mines, jobs that took them all over the United States. Although most were male, women too came to America, albeit in lower numbers. Many Chinese women settled in San Francisco, a city that became the epicenter of the Chinese-American family and, therefore, one in which the battle for public education was fought.[38]

Although other ethnic groups, most notably the Germans, were able to maintain their mother tongues in the public schools of California, bilingual education was largely discouraged among the Chinese. While Chinese immigration was initially encouraged to help the Golden State's economy grow, the Chinese—as more settlers moved west and competition for jobs increased—quickly became white elephants in American society, and being generally unwanted ensured that public education for the Chinese, as with the Native Americans, would be rigidly controlled and highly segregated. The Asian newcomers, the House Committee on Education and Labor noted in 1878, "cannot and will not assimilate with our people, but remain unalterably aliens." As with other ethnicities, white Americans assumed the Chinese were a monolithic group, "invisible," as Ralph Ellison wrote of African Americans, "simply because people refuse to see" them as existential persons. Because the Chinese were an "alien race" (and "invisible" as individuals), political leaders in the late 1870s noted that "education is, therefore, a public necessity, and a move in the nature of self-protection," protection from the "heathenism" of the Asians' "dangerous unarmed invasion of our soil."[39] Not surprisingly, education for the Chinese immigrants focused on Christianity and English, the cornerstones of American civilization.

Because Chinese women and families tended settle in San Francisco, it was the location of the largest concentration of Chinese children. In 1868, for instance, there were some two hundred school-age children in the city; in 1874, there were over one thousand. Some form of formal education, therefore, was needed. In San Francisco and other cities in California, clergymen established a number of Sunday schools and evening schools during the second half of the nineteenth century to "Americanize and Christianize" the newcomers, and many of the pupils were adult men. These religious schools were of the utmost importance because public school officials often neglected Chinese children in the Golden State in general, and in San Francisco in particular. Public education began to take root in San Francisco during the late 1840s and early 1850s, at least for Anglo children. In the mid-1850s, the California Supreme Court determined that the Chinese were not "white," which set the stage for the Asians' racial segregation in education. Beginning in 1859, California established a series of laws that excluded the Chinese from white schools. In the same year, the San Francisco School Board established the "Chinese School," a public, albeit segregated, institution that, during the following decades, was poorly funded and at times closed down. By 1866, the California school law—revised by State Superintendent of Education John Swett—cemented "the legal establishment of separate schools for children other than white children."[40]

Like the schools for American Indians, the Chinese School in San Francisco emphasized English, but its first teacher, Benoni Lanctot, spoke Chinese and presumably bilingually aided his charges' transition to the American language. Moreover, the pupils at the school frequently conversed in their native tongue, which, as with Native-American children, likely resulted in informal, student-led bilingual instruction. In the late 1860s, Superintendent John Pelton—one of the founders of public education in San Francisco—recommended that the Chinese School employ two teachers, one English-speaking and one Chinese-speaking. The superintendent's recommendation came to naught, and the Chinese School kept its official monolingual focus for the remainder of the century. Discriminated against by the *Fon Gwai*, as the Chinese called the racist Americans, the Asian newcomers were largely unable to formally promote their language in the public schools, but the private sector proved more promising for bilingual education. For example, a private Chinese-language school in the city allowed some immigrants to maintain their customs and language. Chinese culture also was preserved through informal discussions and classes in the backrooms of Chinese stores, as well as through Asian theater and the Chinese-English newspapers, such as the *Chinese Record*.[41]

As the Chinese experience demonstrated, the West—that expansive polyglot boardinghouse—was the home of a different sort of public bilingual schooling. Most pronounced among the Native Americans and Chinese, bilingual education was used informally and subversively among solitary teachers and students in order to protect youngsters from the harsh, overwhelming environment found in English-only schools. Because of their non-Christian beliefs, the Native Americans and the Chinese immigrants were subjected to heavy-handed Americanization, a process that aimed at destroying their traditional cultures and languages. Forced into schools to civilize their "savage" and "pagan" ways, American-Indian and Chinese children, along with a handful of compassionate teachers, expressed their ethnic solidarity in English-only schools by informally passing along knowledge bilingually in order to help new students cope with and survive in such arduous classrooms. The public bilingual education in the West before the Progressive era also paralleled the patterns found in other areas of the nation. Germans in parts of Texas and "Spanish Americans" in New Mexico used bilingual instruction to preserve their highbrow and "superior" cultures. In other areas of the West, bilingual education was employed to entice foreign-language speakers into the public schools, which was especially the case for many *tejano* children. Between the 1840s and 1880s, the West, sparsely populated and vast, quickly adopted the long-standing American tradition of localism, and local control ensured that bilingual studies would become a significant part of the public school curriculum in areas with large non-English-speaking populations. Dual-language instruction, therefore, emerged naturally in numerous western localities simply because the schools reflected the linguist particulars of the local community.

5

Nativism among the Homeowners: The Metaphysics of Foreigner-Hating

In his 1857 novel *The Confidence Man*, Herman Melville outlined what he poignantly called "The Metaphysics of Indian-Hating." According to Melville, the hatred of Native Americans originated with the experiences of frontiersmen, experiences that were passed along to the coming generations of backwoodsmen as a sort of cultural lore. Boys on the frontier, were indoctrinated in "histories of Indian lying, Indian theft, Indian double-dealing, Indian fraud and perfidy, Indian want of conscience, [and] Indian blood-thirstiness." Rather than being seen as a vice, the backwoodsmen's Indian-hating was a sort of virtue on the frontier, a virtue that stressed the need for survival and the bringing of "civilization" to the untamed lands of America. In the 1840s, Francis Parkman encountered a fair amount of hatred toward the Native Americans during his travels on the plains. Murdering an Indian in those parts was largely of no consequence, except that such a deed threatened the safety of travelers. Frontiersmen, however, were not the only Americans to utter justifications for their Indian-hating. Such sentiments were also articulated by prominent political leaders, such as Andrew Jackson. One year before he put forth his Indian Removal Bill, President Jackson noted that, despite the efforts of the government, "the Indians in general...have retained their savage habits." For Jackson, the destruction of the Native Americans was almost a mission of necessity because Indians challenged America's moral purity. Indians, it was argued, posed a threat to white women's sexual innocence, white men's Protestant work ethic, and white civilization's realization of a vast agrarian nation. America Indians, however, were not the only recipients of white Protestants' "moral terrorism," a phrase the philosopher Charles Peirce used in 1877 to capture one of the ways Americans ensured conformity in their city upon a hill.[1]

Many immigrant and "non-white" groups in the United States felt the sting of the American majority's moral terrorism because all were considered "foreigners," regardless of their immigration status. While clearly not immigrants, Native Americans were considered by the U.S. government to be "domestic 'foreigners' " because they were ineligible for citizenship. Similarly, Mexican Americans felt as though they were foreigners after the United States colonized the Southwest. After the Anglo influx into Texas, Juan Seguín—once a prominent man in San Antonio—noted he had become "a foreigner in my native land." The designation of "foreigner" was very much tied to language; those who had not mastered the English tongue were foreign, even if they had not emigrated from elsewhere. "A language not understood is gibberish," a German in Baltimore noted, and therefore the speaker "appears dumb, stupid, and ignorant." "Foreigner," therefore, meant much more than "immigrant" during the nineteenth century. The term connoted a person's familiarity with the American language and was used to marginalize certain groups of people. For those who were called foreign, the term became a "hated word," as one German noted, because it suggested that they had "no sympathy with the free institutions of our country" and because it demonstrated that they were somehow different from other Americans. Foreigners, in short, were people to be distrusted and despised.[2] During the nineteenth and early-twentieth centuries, foreigner-hating in the United States was quite pronounced, and this hatred was intimately linked with events in the larger society, as well as with Americans' notions of public virtue.

Public virtue was not static in the nineteenth America; it changed as dramatically as the nation itself. Like the growing American territory, public morality expanded and evolved during the century and mixed with those prejudicial elements that Rogers Smith has argued influenced social policy in the United States. Facilitated by the Second Great Awakening in the 1820s and 1830s, public morality for much of the first half of the nineteenth century was religiously based. Even before the dawning of the nineteenth century, intellectuals such as Benjamin Rush called for a fusion between republicanism and Protestant thought. Rush noted in 1786 that "the only foundation for a...republic is to be laid in Religion. Without this, there can be no virtue, and without virtue there can be no liberty." Well into the nineteenth century, public school textbooks reflected this emphasis on Protestantism as the primary basis for public morality. As the nation grew in wealth during the century, the foundations for public virtue began to evolve, as did Americans' notions of right and wrong. Textbooks, for instance, increasingly focused on secular ethics, rather than Protestant doctrine, as the footing for morality. Reflecting on America's history, H. L. Mencken noted in 1917 that "out of prosperity came a new morality." This new morality—"neo-Puritanism" as Mencken termed it—intimately linked capitalism and the pursuit of wealth with the public good. Of course, theologically based morality did not vanish, but public values were adapting to the material conditions of nineteenth-century America. Perhaps the most dramatic shift in virtue during the century came with the rise of scientism after the Civil War. Science, as well as its accompanying method, became a new moral system for many American intellectuals—a system, however, that did not completely sever its connections with the earlier notions of the public good. But, the growing faith in science did usher in new modes of and justifications for "moral terrorism," particularly "scientific" racialism.[3]

These public virtues greatly influenced the ways in which foreigner-hating manifested itself. Historian John Higham has noted that nativism in the nineteenth and early-twentieth centuries was fostered by a host of shifting demographic, economic, and social conditions, and these changes ushered in three distinct "patterns" of nativist thought, patterns that overlapped considerably. The foreigner-hating of the mid-nineteenth century, for instance, was largely anti-Catholic in nature, while anti-radical and racialist strands of nativism became more pronounced later in the century.[4] As reflections of America's cultural anxieties, these various patterns of foreigner-hating expressed themselves through the moral language of the nation and, therefore, were an indication of the public virtue in the United States. Each manifestation of nativism challenged bilingual education to some extent, and collectively these patterns of foreigner-hating began to unhinge foundational parts of the polyglot boardinghouse.

The American River Ganges

The nineteenth century witnessed significant social, political, and demographic changes in the United States. The opening of the western frontier, the growth of cities and industries, the transportation revolution, the arrival of immigrants, and the rise of popular democracy molded America into something essentially new. Collectively, these changes enhanced nineteenth-century Americans' freedoms. Americans were now freer to move, freer to participate, and, from the diversity of the immigrants, more exposed to ideological and cultural choices. But, with these new freedoms came a certain amount of insecurity and anxiety. Faced with these sometimes frightening social changes and freedoms, many nineteenth-century Americans, particularly the middle classes who adhered to Victorian attitudes, channeled their angst into their child-rearing practices. In the wake of the pan-Protestant Second Great Awakening, Victorians often pressed for a more religiously "rigid" form of upbringing to help ease their insecurities. This form of moral training attempted to give their children a sense of "inner-restraint" in case they happened to encounter the corrupt city, the wicked frontier, or immoral "papist" immigrants.[5]

America was becoming what David Riesman has termed an "inner-directed" society, one that gave young people a "psychological gyroscope" so they would always know how to behave like upstanding Protestants. The moral lessons that children received from their elders emphasized an omnipresent God and moral certitude, lessons that were abundant in nineteenth-century textbooks. One such primer reminded its young readers that children cannot tell half-truths because "God knows *what we mean*, as well as *what we say*." Parents and schools were attempting to make youngsters into what Mark Twain called the "Model Boy," who was "perfect in manners, perfect in dress, perfect in conduct, perfect in filial piety, [and] perfect in exterior godliness." This model boy, Twain noted, "was the admiration of all the mothers, and the detestation of all of their sons." This return to moral rigidity in nineteenth-century America and the making of model children was partially what inspired Mencken to dub such thinking as "neo-Puritanism."[6]

Although neo-Puritans hoped to alleviate their insecurities by turning to a rigid form of moral certitude, this channeling of emotions, especially those as potent as fear and anxiety, could not remain "inner-directed" forever. Unwilling to break with propriety, Victorian Protestants, like their Puritan forefathers, suppressed their darkest emotions, a suppression that "manifested" itself as "a paranoid vision of the outer world." The source of the anxiety—the changes that were bringing new freedoms to Americans—could not be eliminated by individual moral "certainty" alone. Rather, as the eminent social psychologist Erich Fromm has noted of the early Protestants, certitude had to be enforced at the social level as well, leading Fromm to conclude that "[f]reedom and tyranny…were inextricably interwoven." The interconnectedness of freedom and tyranny was also apparent in Protestant America. During his famous visit to the United States in the 1830s, Alexis de Tocqueville claimed, "I know of no country where there is generally less independence of thought and real freedom of debate than in America." Tocqueville attributed this lack of dissent to the "tyranny of the majority," a moral coercion that sought an "egalitarian conformity" to the rigid values of nineteenth-century Americans. The tyranny of the majority was used in America as a "method of authority" for "upholding correct theological and political doctrines," as Peirce noted in the 1870s, and "[c]ruelties always accompany this" method.[7]

After Americans turned to a morally rigid form of pan-Protestantism as an "escape from freedom," the tyranny of the majority manifested itself as a nativism directed against those who did not adhere to those particular religious values. Non-Protestants, of course, were generally considered foreigners, regardless of their citizenship status. As one foreigner-hater noted in 1854, there was a "rapidly increasing bitterness between the Native Americans and the foreigners…[between] the Protestants and the Roman Catholics." The arrival of large numbers of "Romanists" from Ireland and Germany during the first half of the nineteenth century was the primary stimulus for anti-Catholicism in the United States. Fearful Americans saw in this immigration what Thomas Nast depicted in his 1871 cartoon "The American River Ganges," a sketch that showed priests, whose miters were alligator mouths, crawling out of a body of water in order to "devour" good Protestant children.[8] At the time of the foreigners' arrival, the Catholic Church was undergoing a dramatic transformation in Europe, and the immigrants brought with them these new Catholic values. Central to the Catholic "revival" was a commitment to a "communal vision of church, state, and society" a vision that clashed with American individualism and liberal freedom. The revamped Catholic Church also stressed control over its own institutions to insulate its members from heretical views, a move that fueled a certain amount of Protestant paranoia in the United States. "[T]he SECRET POLITICAL ORGANIZATION *known as the* "Catholic Church," one fearful American stated in the 1850s, "*is a perpetual* CONSPIRACY AGAINST THE LIBERTIES OF MANKIND." Together, the communal view and insulated stance generated a general suspicion of Catholics in Protestant America, particularly when these "foreigners" demanded state funding for their schools. In 1840,

Catholics in New York argued that Protestants had a "monopoly" over "public education" and insisted that their schools were "entitled" to a portion of the "Common School Fund."[9] Such demands added to the Anti-Catholic sentiments in the United States and set the stage for the "Know-Nothing" movement in the 1850s.

Religious nativism began to gain strength in the 1830s, but it became particularly pronounced in the 1850s with the rise of the American Party—the so-called "Know-Nothings"—which, according to the German Forty-eighter Henry Boernstein, "was more or less a [nativist] satellite of the Whig Party." The party, with its slogan "America for the Americans," was somewhat secretive about its activities. When members were asked "any questions from a non-member, they were bound to reply, 'I know nothing,'" giving rise to the party's nickname. In an 1854 letter to the editor of the *New York Daily Times*, the writer closed the anti-foreign harangue with "I am content to 'KNOW NOTHING' [b]ut *God, Liberty, and our Country*." The Know-Nothings, as Boernstein reported, "were ostensibly aimed against Catholicism, which was winning ever more property, power and influence. In fact they hated all foreigners without distinction, wanting all public offices to go only to born Americans." German immigrant Carl Berthold noted in 1853 that the "*Wichs* [Whigs] or *Arestokraten* [aristocrats] as they are called in German, and that's what most Americans are, they wanted to treat the Germans unfairly." Ironically, the foreigner-hating of these nativist *Wichs* helped break down Old-World distinctions, allowing an immigrant from the principality of Waldeck to think of himself as "German," rather than solely as a member of a particular region or religion.[10]

Violence was an ever-present aspect of the Know-Nothingism of the 1850s, particularly during election cycles. In 1852, Ned Buntline—"one of the most disreputable nativist rowdies," according to Boernstein—was brought to St. Louis to intimidate foreign voters, particularly those "damned Dutch." After an election day full of brutal fighting, Buntline summed up the events by saying, "Yesterday at the ballot boxes it was Germans and foreigners against Americans!" Three years later, on August 6th, nativist riots broke out in Louisville, during which twenty-two people were killed. Christian Lenz, a German in the city during "Bloody Monday," noted that "these are sad times in America, the Americans are rising up against the Germans with a strong hand." As in Louisville, Know-Nothingism gained strength in Baltimore in the mid-1850s. Many of the nativists in the city were firemen because, as one German noted, "the engine houses...[were] the home of idle and dissolute men." Local fishing clubs also were a repository of foreigner-haters. After the riots following the 1856 election, L. P. Hennighausen commented, "the entire city was in the hands of the lawless element, men, frenzied with bad whiskey,...[who] delight[ed] in shooting and assaulting harmless citizens of German or Irish nativity."[11]

The Know-Nothing movement did achieve some of its anti-Catholic aims, particularly at the local level. The nativist mayor of St. Louis, Washington King, committed himself to "German-eating" and enforcing Sabbath laws. During his rule, Boernstein noted, the "people were held in virtual slavery through restrictions on their freedom. The entire police force became a band of informers, and trial followed trial against German tavernkeepers and against any use of public establishments on Sunday." In the realm of education, Know-Nothingism also had an impact. Anti-Catholicism had long been a prominent theme in many public school textbooks, but the foreigner-hating of the 1850s gave a new urgency to promoting Protestant thought in the schools. In San Francisco, for instance, the Know-Nothing hysteria rekindled the desire to have Bible readings as part of the curriculum; the city's schools had become lax on this issue because of the presence of non-Protestants. Throughout the nation, nativists fought against the use of the Catholic Douay Bible in the common schools and the state funding of parochial schools. In the 1850s, California, with too few schools for its inhabitants, undermined the partial public financing of Catholic schools, initially by requiring parochial school teachers be certified by the state and then by taking control of some of the private schools. Although somewhat successful at leading campaigns against the encroachment of Catholicism in the common schools, the impact of the Know-Nothing movement on public bilingual education was minimal at best. Most of the prominent public bilingual programs, with very few exceptions, gained force after the 1850s, when Know-Nothingism was rapidly declining because of its violence and the coming of the Civil War.[12]

After the war, Know-Nothingism had become a joke in some educational circles. One Illinois school official, after lamenting the lack of knowledge among the state's pupils, sarcastically noted that the students' "obliviousness respecting our country's history would fit them admirably to become candidates for Congress, to be supported by the Know-Nothing party." Although the American Party had died, the hating of non-Protestants had not. During the 1870s and 1880s, religious nativism once again blossomed on American soil, a reawakening that was brought on by a number of sources. At the core of the renewed commitment to the pan-Protestant xenophobia was the arrival of more Catholics in the United States, including large numbers of Jesuits priests who had fled Germany during the *Kulturkampf*. "This country is ruled by priests," one anti-Catholic immigrant noted. "The whole pack of Jesuits that Bismarck chased out of Germany have made themselves at home here." Toward the end of the century, a multitude of Catholics emigrated to America from southern and eastern Europe, thus fueling Protestant paranoia. In addition to the presence of more Catholic and Jewish immigrants, non-Protestants were gaining political power in the last decades of the nineteenth century. Several cities in the East, for instance, elected Catholic mayors during the 1880s. As a response to these alarming trends, several anti-Catholic societies were established in the United States, including the influential American Protective Association.[13]

During the 1870s and 1880s, the schools became one of the central foci of American religious nativism. The primary issues, harking back to the Know-Nothing days, were the public funding of Catholic schools and the use of the Douay Bible in the classroom. As in the 1840s, Catholic Americans saw the public schools as Protestant and, therefore, anti-Catholic institutions, a view that was supported by the continued use of offensive textbooks. In 1890, a Boston school official, Joseph D. Fallon, objected to the use of two high school history books, one of which was "as thoroughly anti-Catholic as the most narrow-minded fanatic could desire." Franz Joseph Löwen lamented to relatives in Germany during the 1880s that "Catholic children...attended the dangerous state schools that we also have to pay taxes for here. Thus the first thing was to start a Catholic school." It was not only Löwen's Michigan community that built parochial schools; such institutions were established all over the nation, particularly after the Catholic Church's Third Plenary Council in 1884. Convening in Baltimore, the council decreed "[t]hat near every church a parish school...is to be built and maintained" and "[t]hat all Catholic parents should be bound to send their children to the parish school." Catholics tried to secure state largess for their schools even before the council's decree for the mass building of parochial schools and were continually offended by the exclusive use of King James Bible in the common schools. These requests outraged pan-Protestant nativists and set in motion the tyranny of the majority to hinder Catholics' demands. Nothing symbolized this campaign against Catholic schooling more than Nast's cartoon in *Harper's Weekly*; the

American school had to be protected from the dangers of the reptilian priests. By the dawning of the twentieth century, the Catholics' desire for state aid had failed. Ironically, the school wars over funding and the Bible not only facilitated the development of more Catholic parochial schools after the Third Plenary Council, but—because of the public debate over the place of religion in the schools—also helped make public education more secular.[14]

This religious strand of American nativism that reemerged in the 1870s and 1880s limited the bilingual activities of some ethnic groups, at least for a time. In the late 1880s, for instance, foreigner-haters in Wisconsin and Illinois targeted the German private schools through the Bennett and Edwards laws, respectively. The two legislative acts, promoted as compulsory education policies, mandated English as the language of instruction in private-sector schools. The laws, however, were bitterly opposed and were quickly repealed. More significant were the nativist sentiments regarding the education of non-European groups. The pan-Protestant mission to Christianize the "heathen" Chinese in California virtually ensured that the immigrants would have very little control over their education, thus limiting their ability to maintain their native tongue in public schools. Like the Chinese, American Indians also had minimal control over their schooling. While some government-supported missionary schools utilized native tongues to educate and "civilize" the Native Americans, fierce battles between Catholics and Protestants in the last few decades of the nineteenth century undermined the use of these sectarian "contract" schools. Disturbed by the Catholic Indian schools receiving federal funds, the Protestants affiliated with the Friends of the Indian at Lake Mohonk and the American Protective Association supported a shifting of all educational responsibility to the U.S. government, which by the end of the nineteenth century staunchly supported English-only instruction. Although the government monopoly over Indian education would end the Protestant contract schools as well as those of the Catholics, Protestants were willing to forego their educational institutions to end the papist threat. Moreover, foreigner-haters in the latter part of the nineteenth century were convinced that the "Romanists" were opposed to the English language. Fueling this belief was a forged Catholic document that circulated in nativist circles. The supposedly official decree stated that "[w]e view with alarm the rapid diffusion of the English language. It stands before the world as the tongue which has for 300 years ever been opposed to our holy church." The activities of the Lake Mohonk group and the American Protective

Association were largely successful. Not only had government funds to the Catholic missions been severely cut by the turn of the twentieth century, but also, with the sectarian schools on the wane, English had become the only legitimate language of instruction in the Native American schools.[15]

Survival of the Fittest

By all accounts, nineteenth-century Americans were a hard-working lot, full of "restless, nervous energy," as Frederick Jackson Turner put it. In the 1830s, Tocqueville noticed that the inhabitants of the United States were fanatically busy with working and making profits. He stated that "[a]ll constantly wish to acquire material possessions, reputation and power," although very "few have a lofty conception of these things." Tocqueville's description of the hard-working, money-hungry American was repeated by Melville at mid-century. Many of Melville's characters in *The Confidence Man* were small-scale capitalists, hoping to earn a lucrative return on their hard-earned dollars through investment; the "confidence man," the profit-seeking con-artist, was always more than happy to oblige others' greed with some phony company in which to invest. Reflecting on the American of the nineteenth century, historian Henry Steele Commager has noted, "To the disgust of Europeans, who lived so much in the past, he lived in the future, caring little for what the day might bring but much for the dreams—and profits—of the morrow." All of this frenzied work sometimes prevented Americans from carefully reflecting upon their anxieties and passions regarding important social issues because "Americans [were] too busy making money to analyze" such matters.[16]

Protestantism and capitalism complemented each other fairly well because both stressed the benefits of industriousness, and, because of the connection between the two, capitalism came to be seen as its own virtue. Discussing the 1920s, media scholars Stuart Ewen and Elizabeth Ewen have argued that "free-market capitalism...was elevated to the status of an unassailable secular religion."[17] The linking of free-market economics with virtue in the United States had its roots in the nineteenth century, however. This rise of the capitalist aspects of American public morality even became significant with regard to the most important moral issue of the century: slavery. Until the 1840s, the bulk of the abolitionists' criticism of the Southerners' "peculiar institution" took the form of traditional moralizing, particularly by

pointing out the evils of such a system and how it violated both natural and divine law. After the 1840s, a new moral criticism of slavery
began to take center stage, a criticism that was economic in nature.
Abolitionists argued that slavery was an evil system because it undermined capitalism and the Protestant work ethic in the South. The economic aspects of Protestant morality grew in strength after the ending
of slavery. It was, after all, decades after the Civil War when Mencken
claimed that the United States was participating in "neo-Puritanism,"
a revivalist Protestant movement that had grown powerful with capitalist wealth. "The [new] American Puritan," Mencken noted, "was
not content with the rescue of his own soul; he felt an irresistible
impulse to hand salvation on,…to ram it down reluctant throats"
because now "he had the money" to do so.[18]

Not only was capitalism morally right for many nineteenth-century
Americans, it was also natural. The capitalist system was "survival
of the fittest" in the economic realm, as William Graham Sumner
described it in 1881. It was a system not "made by man," rather one
that emanated from the natural order of the universe. Therefore,
socialists and all others who challenged capitalism were not only evil
because they disturbed the free-market orientation of the United States,
but also because they violated natural, and therefore—presumably—
divine, law. Leftist European immigrants, naturally, would have to
contend with the tyranny of the majority. Some European newcomers
were a source of economic uneasiness even before the Civil War, such
as the radical Forty-eighters. In addition to the arrival of the political
exiles of the European revolutions, 1848 was also the year in which
Karl Marx and Frederick Engels' *The Communist Manifesto* was
published. The pamphlet held within it ideas that threatened the very
foundations of the American economy. Marx and Engels concluded
by stating that "Communists…. openly declare that their ends can be
attained only by the forcible overthrow of all existing social conditions. Let the ruling classes tremble at the Communist revolution. The
proletarians have nothing to lose but their chains."[19]

Tremble they did, but not until the last quarter of the nineteenth
century when radicalism in the United States appeared to be on
the rise, thus causing a surge in economically based foreigner-hating. International events, particularly the emergence of the Paris
Commune in 1871, fueled a general fear of radicalism. Although
the Paris Commune was short-lived, the anxiety over the rising up
of workers resonated across the Atlantic. During the 1870s, many
Americans were becoming suspicious of leftist Europeans, fearing

that they might come to the United States filled with Marx and Engels' seditious ideas and stir up labor unrest. Suspicion begat paranoia, which dramatically increased after the Haymarket bombing in 1886. The anti-radical paranoia ushered in a wave of nativism that sought to exclude and deport foreign leftists from the United States. After President William McKinley's assassination by a radical in 1901, immigrants' political and economic leanings were carefully scrutinized before being allowed into the city upon a hill.[20]

In addition to immigration restriction, Americans also sought to ensure conformity to their economic morality though their institutions. The public schools actively attempted to indoctrinate capitalist values, particularly the values of the industrious middle classes. The nineteenth-century textbooks directed at the mass of American students often stressed the virtues of hard work, industry, and thrift. The neo-Puritans' efforts to instill the Protestant work ethic and faith in capitalism were perhaps most pronounced with regard to the "hated": African Americans and foreigners. Schools such as Hampton and Tuskegee, with their focus on "industrial" training, became the model for black education in the South. Students in these schools learned the morality of capitalism through practice and were thus fitted for occupations on the lowest rungs of the free-market hierarchy. Many newly arrived immigrant children learned similar lessons in their public schools, lessons that helped them become "efficient" factory workers for the nation's great industrialists. Paralleling Woodrow Wilson's view of the people colonized after the Spanish-American War, the "hated" in the United Stated first had to learn "discipline" and "order" before they could be trusted to be full members of America's capitalist society.[21]

While the local public schools were attempting to thwart radicalism by instilling in immigrant students the tenets of the free market, many of those who failed to worship at the feet of capitalism were not newcomers. Many American Indians—who, as the Board of Indian Commissioners noted in 1881, were considered "foreigners"—had a communal, almost socialistic lifestyle that did not go unnoticed by federal officials. A "tribe, with its territory held in common," noted the secretary of the interior in 1879, was not living the American way. The ending of this threat required the fostering of "individual ownership of land," as well as education. The Dawes Act of 1887, which began the process of breaking up of tribal lands into single-family plots, was an attempt at promoting the capitalist spirit among the American Indians. But, education was still a necessity, an education

that cultivated the Protestant work ethic among the rising generations. The Hampton formula for educating former slaves was applied
to American Indians, and some even attended that institution before
the Carlisle Indian School opened in 1879. The English-only, off-
reservation boarding school, like Carlisle, became the federal government's favored mode of Indian education. Indian children, the
superintendent of Indian schools stated in 1887, were best "reached
through a boarding school" where they learned "to speak the English
language" and received an "industrial training...[that] will cultivate
the habits of industry and thrift," habits that prepared them for their
"allotment of land...as they reach the proper age."[22] Such policies
intended to end the socialism among America's "domestic foreigners"
by inculcating the virtues of capitalism.

While the fear of radicalism was fueling a great deal of the nativist
feeling, the cycles of the capitalist system were also generating foreigner-hating in the United States. In California, for instance, the
newly arrived Chinese immigrants were, at least initially, greeted
with open arms; they had come to work and build the Golden State's
economy. But anti-Chinese sentiment grew throughout the 1850s,
coinciding closely with an economic decline in the state. John Swett,
a San Francisco schoolteacher and state superintendent in the early
1860s, remembered that "in 1854 there were heavy business failures"
in San Francisco, "and in 1855 a great financial collapse began. There
were one hundred and twenty-seven failures, amounting in all to
eight millions of dollars." With anti-Chinese feelings on the rise, the
immigrants' clout waned, thus leaving them with little power to cultivate their native tongue in public institutions. The Chinese were not
the only immigrant group to feel the wrath of capitalist cycles. The
Norwegians also experienced setbacks from economic downturns.
The depressions in the 1870s and 1890s slowed immigration from
Scandinavian lands considerably.[23] Because local political power
often came with large numbers—as the Germans demonstrated—
fewer countrymen meant less influence and, therefore, a hampered
ability to establish more Norwegian-English schools in the regions
they settled.

Economic problems also threatened the most secure of the public
bilingual programs in the nation, those of the German Americans.
Historian Steven Schlossman has correctly noted that "German
instruction was generally most vulnerable to attack during periods of
nationwide financial distress, as during the depressions of the 1870s
and 1890s." In the 1870s, states such as Kansas and Ohio revised their

school laws in order to restrict bilingual education. Kansas ended all German-language instruction in 1874, while the Ohio legislature, during the previous year, attempted to eliminate German as the language of instruction in many of the state's public schools by mandating that *Deutsch* be taught as a subject only. In addition to the states, local communities also began to restrict German-English schooling. Throughout the 1870s, attacks on German-language instruction occurred in St. Louis and Milwaukee, although, at least temporarily, the German communities in those localities were able to fend off the criticisms. During economic downturns, aspects of the public schools that were seen as "foreign" were considered unnecessary or un-American and, therefore, were earmarked for elimination.[24]

The 1880s and 1890s witnessed the continuation of the fiscal attacks on German-English instruction, and the challenges in these decades proved to be more formidable. The battles over the German language in the public schools of St. Louis, for instance, did not subside with the depression of the 1870s. Although milder than in the decade before, depressions continued to crop up during the 1880s, as did the attacks on the city's bilingual program. In 1887, the president of the school board reported that it had been necessary to hire more teachers who could teach both languages—instead of just special German teachers—"as a compromise measure, sanctioned for economy's sake." In January of the following year, German-language instruction was eliminated in the city. The fiscal criticisms of German-English schooling began in Indianapolis during the mid-1880s, culminating with the temporary ending of the bilingual program in 1890. Although, after public outcries and a legal battle, German-instruction continued in Indianapolis, the 1890s were hard on German-English bilingual education. The depression was overwhelming and quite incomprehensible. In 1892, German immigrant Michael Probstfeld wrote that "it does seem strange to me that when the granaries burst from overloading, when heaps of stored clothing are being eaten up by moths,—there are so many hungry and naked." "There's a screw loose somewhere," Probstfeld continued, "[b]ut who is the mechanic who knows which one?" It was within this context of the economic distress of the 1890s that both Chicago and New York revised their policies regarding German-language instruction. In 1893, Chicago's public schools pushed back the beginning year of German studies from grade three to grade five. Four years later, New York City's schools moved foreign-language instruction to grade six; previously it had begun in grade three. The financial troubles of the late-nineteenth century were not the only reason for the attacks on bilingual education. Political and ethnic rivalries also played a role, as did the rise of a racialist form of nativism, a nativism that took root as new groups of immigrants began arriving from southern and eastern Europe.[25]

The Great Chain of Being

By the 1870s, moral certitude in America—as well as the Victorian sentiments from which it flowed—was increasingly being challenged in intellectual circles. Louis Menand has argued that the breakdown of nineteenth-century morality was a direct result of the war, noting that "the Civil War discredited the beliefs and assumptions of the era that preceded it. Those beliefs had not prevented the country from going to war...[or] prepared it for the astonishing violence the war unleashed." As a soldier in the Union Army, Oliver Wendell Holmes, Jr. learned a great deal from the war. In particular, he came to understand that "certitude leads to violence." Having learned that important lesson from the conflict, Holmes's later legal scholarship reflected his criticism of "[c]ertitude" and "natural law" and, therefore, argued for a more contextual view of issues. The Civil War was not the only element that was gradually changing the ethos of post-bellum America. A growing faith in science also led many intellectuals to question moral rigidity during the last decades of the nineteenth century. Science was reshaping the Western world, giving inspiration to public intellectuals who would eventually unravel Victorianism even in its birthplace of England. In the United States, philosophers such as William James and the other pragmatists drew on science and its method to analyze morality. The answers that the pragmatists brought to moral issues questioned the notion of absolute "truth" and helped usher in value systems among some intellectuals that were more flexible than those that had dominated American thought prior to the Progressive era. "[T]here can be no final truth in ethics," James noted in 1891, "until the last man has had his experience and said his say."[26]

While the rise of scientism was partly responsible for the chipping away at moral certainty in the last third of the nineteenth century, many American intellectuals, interestingly, turned to science for moral direction. The scientific method, for example, was sometimes used as an ethical system in universities, one that provided guidance for solving problems and seeking truth. In the 1880s, David Starr Jordan, president of Indiana University, hoped to substitute the institution's Protestantism and *in loco parentis* duties with a devotion to science and scholarship. This change, Jordan claimed, would give the students "higher ideals, [and] more serious views of life." In some circles, therefore, science was becoming intimately linked with morality. "The ultimate end of science, as well as its initial impulse," Jordan noted after he went to Stanford, "is the regulation of human conduct. To make right action possible and prevalent is the function of science."[27] For some, science could be a substitute for rigid Protestantism.

America's moral views helped define the nature of nativism, and as public virtue became more "scientifically" oriented, so too did foreigner-hating. During the late-nineteenth and early-twentieth centuries nativists increasingly drew upon racial "science" as the basis for their xenophobia. In the United States, racialist sentiments had long been applied to non-European groups: Blacks, Indians, Latinos, and the Chinese. By the Progressive era, racialism—bolstered by the rise of scientism—was being used to marginalize certain groups of Europeans as well, particularly the newcomers from the southern and eastern sections of the continent. This racial thinking, therefore, was "scientifically" subdividing the links on the "Great Chain of Being," the notion that a hierarchy existed between the lowest life forms and God, and, of course, the people who did not emigrate from northern or western Europe were placed on the lower rungs of the hierarchy of creation. This classification of peoples into higher and lower races, naturally, denied the humanity of those deemed inferior by ignoring their common bond to the larger community of mankind.[28]

Racial science had been developing for centuries; as Europeans had more contact with the peoples of the world, racism was used as a means to justify colonialism and slavery. But, Americans also contributed to this growing field of study, particularly in the nineteenth century. As a follower of Herbert Spencer, Sumner's sociological work promoted social Darwinism in the United States, particularly by endorsing the view that "[n]ature .... grants her rewards to the fittest ... without regard to other considerations of any kind" and that "the social order is fixed by laws of nature." Much of the racial science in late-nineteenth century America was a result of the arrival of vast numbers of new Europeans groups. In the 1890s, nearly two million southern and eastern Europeans emigrated to the United States, many of whom were Czechs, Poles, Hungarians, Russians and Italians (see table 4). During the following decade over six million

Table 4 Number of Eastern and Southern European
Immigrants by Decade

Immigrant Group	1880s	1890s
Czechs (Bohemians)	33,000	39,000
Poles	99,000	236,000
Hungarians	51,000	83,000
Russians	147,000	241,000
Italians	138,000	301,000

Source: U.S. Immigration Commission, *Statistical Review of
Immigration, 1820–1910—Distribution of Immigrants, 1850–1900*
(Washington, D.C.: Government Printing Office, 1911), 417.

arrived, with the largest number coming from Italy. The arrival
of different foreigners encouraged social scientists to examine the
racial implications of the "new" immigration on America and its
institutions.[29]

Many of those who began to think of European immigrants in
racialist terms drew on anthropology, a science that, in many ways,
was born from Europeans' fascination with human inequalities.
Daniel Brinton, a leading anthropologist in the late-nineteenth century, argued that there was a racial hierarchy of human beings. The
" 'lower' race[s]," Brinton stated, had much in common with apes.
Drawing on Brinton's work, the U.S. Immigration Commission outlined the "racial" characteristics of people in the early decades of the
twentieth century. The commission's report suggested that Anglo and
Germanic peoples had numerous positive attributes, while the southern Italians, the largest group of newcomers at the time, were prone
to criminal activity and had "little adaptability to highly organized
society." This revised form of foreigner-hating essentially posited
that the "new" immigrants from southern and eastern Europe were,
by their very nature, lower on the Chain of Being. With such a bio-
logically determined stance, racial nativists feared that assimilation
to American society would be difficult, if not impossible, for some
immigrant groups because of their "inferior" make-up. Without the
faith that the environment could greatly shape and improve people—a
view more common in the 1840s—many social Darwinists increasingly called for a restriction on the number and type of immigrants
coming to America.[30]

Yet, the policies of the decades surrounding the turn of the
twentieth century were multifaceted and, at times, contradictory

regarding immigrants. Although foreigner-hating social Darwinists emphasized biological determinism and, therefore, wanted immigration restriction, Darwinian thought also highlighted the power of the environment in the shaping of organisms. The social Darwinists who claimed that heredity was the most important factor for natural selection were a vocal lot, but the environmentalists were a powerful force in intellectual circles during the Progressive era. Lester Frank Ward, the eminent American social scientist, insisted that humans had the capacity to change, depending on their social settings. For Ward, biological determinism was mistaken because it overlooked the power of the human mind in the shaping of natural world. "Nature," Ward noted in 1884, "has thus been made the servant of man." If the mind and the environment were also factors in the shaping of human being, as intellectuals such as Ward, Percy Stickney Grant, and Franz Boas argued, then education was of the utmost importance.[31]

Fearful of the new immigration, many Progressive-era Americans wanted the foreigners to assimilate rapidly, and the public schools were called into action for this mission. The school was supposed to be one of the central ways in which the United States lived up to its melting-pot ideal, but the ideal was losing its luster with the arrival of unwanted foreigners. In Israel Zangwill's play, *The Melting Pot*, the central character stated that "America is God's Crucible, the great Melting-Pot where all the races of Europe are melting and re-forming." But, the "Melting Pot," as Robert Wiebe has noted, was "always producing the same ore from any mix of ingredients." Many Progressive-era schools were engaged a "smelting process," as historian Stephan F. Brumberg has poignantly argued, "creating [an] intense heat that would burn away the dross of alien cultural impurities." In addition to the assimilation goal, the public schools, always a reflection of the larger society, were permeated with racialist thought in other ways as well. Textbooks, for instance, mirrored the focus on race. While anti-Catholic sentiments pervaded texts for much of the nineteenth century, "[b]y the end of the century," one scholar has noted, "questions of race and nationality occupy the schoolbooks more than does that of religion." At the turn of the twentieth century, Adele Marie Shaw studied New York City's public schools and found that many teachers viewed their immigrant students through a racial lens and required strict obedience from their "dirty little Russian Jew" pupils. As a major destination for the new immigration, Shaw understood that New York and its schools were doing much to assimilate the newcomers, but the task was overwhelming; she, therefore, concluded that immigration restriction was needed.[32]

Racialist attitudes during the Progressive era intensified the smelting function of the public education system, particularly by its emphasis on the Americanization of foreigners. Although the Americanization of immigrants did not coalesce into a full-blown movement until World War I, the movement's roots were growing prior to the international conflict. The public schools encouraged foreign children, particularly those from southern and eastern Europe, to shed their cultural heritages and to adopt America's language and customs. "Prior to 1880," the superintendent of New York's public schools, William L. Ettinger, announced in 1918, "the tide of immigration flowed from northern Europe and meant we were recipients of stock...who were easily assimilated because of certain common bonds." "However," Superintendent Ettinger continued, "since that time the immigration movement has swept southward through Europe and has brought to our shores people whose racial history makes the problem of Americanization or assimilation much more difficult." Regardless of the difficulties, Ettinger was confident that assimilation could be achieved through careful training in English and civics.[33]

The racialist strand of nativism that helped give rise to the intensification of the assimilation goal naturally worked against the development of public bilingual schooling among the new immigrants. While the new conformist mood of the nation was hampering the growth of bilingual instruction, the southern and eastern Europeans did achieve some scattered successes in maintaining their mother tongues in schools. The Poles, for instance, established numerous parochial schools in which Polish was the language of instruction. These schools often became bilingual institutions as the Polish community and, especially, the clergy and teachers gained more familiarity with the English language. In rural enclaves in Texas, the Czechs made their native tongue part of the curriculum in both private and public schools. Keeping with the tenor of the times, Czech leaders argued that the language would foster assimilation by enticing foreign children into the bilingual public schools. During the early twentieth century, Milwaukee's public schools offered both Polish and Italian to its students. In 1916, over three thousand pupils studied Polish in nine elementary schools throughout the city, while just over nine hundred students learned Italian at the Detroit and Jefferson Street Schools.[34]

By and large, however, the focus on assimilation for the new immigrants made public bilingual schooling a difficult endeavor, and it threatened the bilingual programs of the more-established groups of foreign-language speakers as well. New York City, with its large portion of foreigners from southern and eastern Europe, initially allowed its new foreign-language-speaking students to simply "sink or swim" in its public schools. Without any bilingual programs for the newcomers, all students with no knowledge of the English language—the Italians in the sixth and fourteenth wards and the Jews on the Lower East Side—were placed in the first grade, even older students. By the early twentieth century, the city's public schools had developed special classes, the so-called "C classes," to help Jewish students make a quick transition to English. Although these courses did not utilize the students' mother tongue, which made them more akin to today's English as a Second Language (ESL) programs, they at least recognized immigrant students' special linguistic needs. In addition to the "C classes," New York City's public schools also held evening studies to facilitate the learning of English. Grouped by age and native language, students in the evening courses learned English words through the teacher's demonstration of verbs and illustrations of nouns. This focus on English in the city's schools undermined foreign-language instruction, however. By the early-twentieth century, German and French became elective subjects beginning in the eighth grade. As in New York City, foreign-language instruction was on the defensive throughout the nation during the Progressive era. The attacks on bilingual education in the 1880s and 1890s were not solely because of fiscal issues; the racialist strand of foreigner-hating and the Americanization efforts it brought forth also contributed to the challenges to bilingual instruction in the public schools of the United States.[35]

The views and attitudes that ushered in the racial strand of foreigner-hating, particularly scientism, were reshaping America during the Progressive era and challenging the Victorian morality of the nineteenth century. Many Anglo Americans, such as Prescott F. Hall of the Immigration Restriction League, had a paranoid view of the United States' future, a view that suggested that the "barbarians" from southern and eastern Europe would upset the racial balance of the nation through intermarriage. Increasingly, Anglos, with a lower birth rate than the newcomers, began to worry that "race suicide" was a real possibility, particularly after the publication of Madison Grant's *Passing of the Great Race* in 1916. The prudishness of

Victorian morality also merged with modern science. Many, like the sociologist Edward Ross, believed that immigrant groups had particular racial characteristics that led to specific vices, such as "sexual immorality" and intemperance. New social agencies, making use of the emerging sciences, attempted to regulate these immoral activities by closely supervising immigrant families.[36] The efficient regulation of society became one of the hallmarks of progressivism in America, and certain aspects of "progressive" thought, particularly in education, sought to undo the polyglot boardinghouse.

6

Progressivism, Science, and the Remodeling of the Boardinghouse

While nativism was chipping away at the polyglot boardinghouse, the coming of progressivism—as well as the new social sciences that bolstered it—was rocking the foundations of localism that supported American multilingualism. By the late-nineteenth century, the nation was undergoing a dramatic transformation, a transformation that essentially ushered in a new society. Reflecting on the 1890s, Theodore Dreiser commented that the "spirit of America at that time was so remarkable. It was just entering on that vast, splendid, most lawless and most savage period in which the great financiers... were plotting and conniving the enslavement of the people and belaboring each other for power." Progressive presidents, such as Theodore Roosevelt, attempted to end the most egregious aspects of that corporate lawlessness by asserting the power of the federal government. "Corporations," Roosevelt told Congress in 1901, "should be regulated if they are found to exercise a license working to the public injury." The goal of many progressives was not just to put a stop the lawless and savage ways of big business. Their ultimate aim was to build a more unified and cohesive nation, an aim that required the replacing of America's "island communities," as Robert Wiebe has called them, with a strong, centralized government. Roosevelt hoped to achieve that centralization through his New Nationalism when he made a failed bid for the presidency in 1912. "The New Nationalism," the ex-president explained in 1910, "puts the national need before sectional or personal advantage.... [and] regards the executive power as the steward of the public welfare."[1]

The drive for a more cohesive society partially reflected an anxiety about an America that had grown out of control. During the late-nineteenth century, industrial development surged to new heights, and not just in the nation's premier cities. Smaller urban centers, such as Milwaukee, Toledo, and Rochester, witnessed booms in industry that must have seemed unimaginable just a few decades earlier. Despite intermittent depressions, the banking and financial sectors also blossomed in the decades surrounding the turn of the twentieth century. With men such as J. P. Morgan at the helm, America's financiers had the capital to prod industrial development in the United States, thus making the nation less dependent on foreign credit. The railroad industry, which had long been fingered as a source of unchecked power, continued to grow, adding four transcontinental lines after the depression of the 1870s. Although America was rife with technological advances, these modern innovations, from the perspective of many, were causing the nation to spiral out of control. Modern "[m]an," Sigmund Freud wrote, "has become a kind of prosthetic God. When he puts on all his auxiliary organs he is truly magnificent; but those organs have not grown on to him and they still give him much trouble at times."[2] The Progressive era was certainly one of those times in which modernity was giving the prosthetic God trouble.

During the late-nineteenth century, the economic sector was not the only source of distress. The immigrants from southern and eastern Europe arrived at an alarming rate, and the older, more-established foreigners were increasingly asserting their power in the political realm. In 1898, John Fisk, intellectual and president of the Immigration Restriction League, told William Lloyd Garrison that Americans "have a perfect right to exclude persons whom we think for any reason would be undesirable citizens."[3] These anxieties about immigrants and cohesion, in an age of scientism, could be addressed only in the most systematic of ways. The polyglot boardinghouse, long sustained by apathy and localism, was shaken by the winds of centralization, standardization, and foreigner-hating. Not surprisingly, many public bilingual programs became causalities during the progressive campaign for unity, but others flourished. Like the larger society, bilingual education was in flux. New justifications came to replace the old notions of cultural maintenance, justifications that drew from the scientific and reform-minded sentiments of the age and, ultimately, remodeled the polyglot boardinghouse.

A Progressive Society

Progressivism was a variegated and multidimensional movement that began to take shape during the late-nineteenth century. The "heart of Progressivism," historian Morton Keller has insightfully noted, was "a shared sense that the good society was efficient, organized, [and] cohesive." Coinciding with the emergence of a modern capitalistic society, progressive thought, at times, merged with the American dream of wealth. Writing on the trends in the late-nineteenth century, Dreiser commented that "[t]here was almost an angry dissatisfaction with inefficiency, or slowness, or age, or anything which did not tend directly to the accumulation of riches." But, many progressives railed against the unfair accumulation of wealth, particularly through exploitation, trusts, and monopolies. "Progressivism," therefore, was an all-encompassing concept "which," as Richard Hofstadter has put it, "for the lack of a better term might be called 'activism.' " As a movement of activism, progressive leaders spanned the political gamut. Therefore, the Socialist Eugene V. Debs, the Social Gospeler Walter Rauschenbusch, the labor leader Samuel Gompers, and the conservative sociologist Edward A. Ross were all, in different ways, participating in a "progressive" movement to bring order to America's evolving, increasingly impersonal society.[4]

The drive to make the United States a more efficient, organized, and cohesive society stemmed from a number of sources, but perhaps the most salient was that America in the last decades of the nineteenth century was becoming utterly complex. From the perspective of many, the social and economic changes that characterized the Gilded Age were baffling. Powerful railroad companies increasingly shaped the lives of the Americans who relied upon them, most notably those involved in agriculture. Some of the early expressions of activism, therefore, came in the form of regulating transportation. The Granger movement in the Midwest and Populism in the West fought the dominance of the railroad companies. Currency too had become complicated, and the debates over what form of specie, whether gold or silver, ought to be the standard was an attempt to bring some sense of order to the increasingly complex economy. While influential corporations and the monetary system were difficult to comprehend, perhaps nothing was more unsettling than the series of depressions that shook the nation in the late-nineteenth century. As a farmer in Minnesota, Michael Probstfeld could not understand nor escape the onset of national economic problems in the 1890s. Although the granaries were full, Probstfeld confusedly commented, "[t]here are still people...going without bread...[because they] are too poor to buy it." Probstfeld's friend, the prominent Forty-eighter Albert Wolf, committed suicide in 1893 because of the financial strain of the economic crisis. The depression was incomprehensible because it undermined central American beliefs, beliefs such as the Horatio Alger quip that "your future position depends mainly upon yourself,...it will be [as] high or low as you choose to make it." Economic recessions demonstrated that such perspectives were not entirely true. Americans did not choose to go without bread, and "social observers," as William J. Reese has noted, "agreed that suffering was widespread and too great to be explained by personal deficiencies." The depression—baffling like other aspects of the modern economic system—unleashed chaos, a chaos that witnessed, as Probstfeld noted in 1894, "bank after bank...going bankrupt, [as well as] business firms, factories, [and] fire and life insurance companies."[5] In this age of uncertainty, Americans increasingly demanded stability.

Perhaps no single group sought out social stability more than the "new" middle class. As the "island communities" continued to break down in the late-nineteenth century, some Americans found a substitute cohesion in their social class. Professionals, skilled laborers, and others in the middling groups began to unite through associations. Through their specialized organizations, teachers, physicians, attorneys, and artisans came to share a worldview, a worldview that emphasized social unity in uncertain times. For professionals, the key to unity was found in the emerging social sciences, which, under the guidance of experts, could ultimately cure the ills of modern America. Drawing on their positivist faith, many middle-class Americans felt confident that science, when applied to social problems, would usher in a standardization and efficiency that would bring order to the nation's confusion. The scientific order the new middle class sought was often bureaucratic. "Bureaucratic thought," Wiebe has noted, "made 'science' practically synonymous with 'scientific method.' Science had become a procedure, or an orientation, rather than a body of results."[6]

With their faith in bureaucratic procedure, the members of the new middle class were ushering in a different vision of the model American, one who had a different set of values from his or her counterpart at mid-century. The United States had become impersonal, complex, and ever-evolving, which necessitated, according the middle-class promoters of the new American, not only bureaucratic management, but also a certain amount of flexibility in order to adjust to the changing times. The nation, in fact, had emerged as an "other-directed" society, as David Riesman has called it, a society in which *contemporaries are the source of [ethical] direction for the individual."* George F. Babbitt, Sinclair Lewis's famous caricature of the American businessman, embodied the new middle-class values. Babbitt's morality came from his peers and changed with the situation. Moreover, he was infinitely practical, "efficient, up-to-the-minute and...modern." But, ethical flexibility required the modern American to have a good "personality," presumably to demonstrate his or her adherence to middle-class mores. Babbitt and his contemporaries, for instance, recognized the "Pulling Power that Personality" had in the business world. Dreiser too noted that "any man who does anything must have so much more than the mere idea—must have vision, the ability to control and to organize men, a magnetism for those who are successful—in short, that mysterious something which we call personality."[7]

Although personality was essential in modern America, it was not the key to progressive thought. "The heart of progressivism," Wiebe has written, "was the ambition of the new middle class to fulfill its destiny through bureaucratic means." This destiny was reform or, as the sociologist Lester Frank Ward termed it in 1884, "a truly *positive* philosophy, i.e., a philosophy of *action."* Many members of the middle classes were moved to "action" during the decades surrounding the turn of the twentieth century because the problems associated with modern, industrial, and urban life were brought before the public's eyes. In 1890, Jacob Riis's *How the Other Half Lives* exposed the terrible conditions of tenement life in New York City. Upton Sinclair's 1906 *The Jungle* also brought to light the horrors of the tenements, as well as child labor, but the book primarily focused on Chicago's meatpacking industry. Of the odor of the stockyards, Sinclair wrote, "you could literally taste it, as well as smell it—you could take hold of it, almost, and examine it at your leisure." While shoveling "guts" and working on the "killing beds," employees witnessed the unsanitary conditions of meatpacking; processing rotten meat was not uncommon, and for sausage, everything from the floor went into the hoppers: sawdust, "consumption germs," dirt, rat poison, and vermin. Of the largely immigrant workers, Sinclair commented, "[t]his was in truth not living; it was scarcely even existing." In addition to the book-length exposes, the widely circulating magazine *McClure's* was an outlet for the so-called muckraking journalists, including Ida M. Tarbell who examined the monopolistic practices of Standard Oil in 1902.[8]

Progressive reformers, driven by the middle-class desire for bureaucratic order, achieved many gains at all levels of government during the decades surrounding the turn of the twentieth century. The Pure Food and Drug Act in 1906 sought to end the sort of problems that Sinclair described in *The Jungle*. The Hepburn Act gave more power to the Interstate Commerce Commission, while the Sherman Anti-Trust Act helped undermine the ubiquitous power of the railroads. Much of the thrust of progressive reform was for a more-efficient, less-corrupt, and activist form of governance because by the turn of the twentieth century, as Sinclair noted, "humanity was festering and stewing and wallowing in its own corruption." Local and state officials, therefore, entered into battle with the political machines and bosses. At the federal level, opposition to the corrupt spoils system generated a movement for an expanded civil service, which, it was believed, would make the nation run more efficiently and, to some extent, without the tarnish of political machinations. These reforms, however, were often laid on top of the existing governmental structure—which continued inefficiency to some degree—and often created unforeseen consequences, such as departmental politics. Local, state, and national progressive leaders also did much to regulate child labor. After all, Wiebe has noted, the "central theme" of the humanitarian form of progressivism "was the child."[9] This progressive drive to save the child facilitated the push for new forms of public education in the United States.

A Progressive School

Throughout the nineteenth century there had been numerous campaigns to "save" America's children, campaigns primarily directed at poor and immigrant youngsters who were dependent, neglected, orphaned, or delinquent. In tune with the activist mood of the nation, child saving intensified during the Progressive era, and there were plenty of ills from which to save youngsters during the period. Because of the density of tenement life, urban children were often exposed to communicable diseases, as well as the ubiquitous "vermin." Augie March, Saul Bellow's famous character, encountered "rat-bitten kids" when he became a house inspector in Chicago. Moreover, children participated in many dangerous activities, such as playing in traffic-filled city streets which, as Sinclair noted, contained "stinking green water." Youngsters explored the urban dumps and rubble, an activity which, although hazardous, was not only entertaining, it also, as the

Italian immigrant Leonard Covello noted, could be quite lucrative if one happened upon something worth selling. Many of these city children worked long hours as street peddlers or in factories and tenement sweatshops, and with their earnings, some moved out of the family home at an early age. Domestic work within families often fell upon the eldest daughter, which, as the Jewish immigrant Mary Antin noted of her sister, made for "a joyless childhood." As in the cities, girls in the rural areas labored endlessly, which sometimes prevented them from attending school. Ántonia Shimerda, Willa Cather's Bohemian heroine, noted that "[s]chool is all right for little boys. I help make this land one good farm." By 1900, conservative estimates suggested that there were nearly ten million children between the ages of ten and fifteen in the American workforce.[10]

The campaigns to protect America's children were often led by women. Stereotyped as the "Gibson girl"—from Charles Dana Gibson's drawings—the "new" woman, often a product of the middle and upper classes, increasingly expanded her role in the public realm by broadening the sphere of domesticity beyond the home. In 1898, Charlotte Perkins Gilman noted that women had long been held in virtual "slavery." Under the thumb of men and chained to the home, America's new women made the case that public activism on behalf of children was a logical extension of their traditional duties. Jane Addams, for instance, argued that a college-educated woman needed an outlet for the desire to be "a citizen of the world." Settlement work, which often focused on the needs of children, was just the sort of "health-giving activity" that the new woman needed because it "formulate[d] into action" the ethical principles she learned in college. These female reformers were in the forefront of progressive campaigns for improving children's health and creating playgrounds, as well as professionalizing the social work that took up such causes. After the 1909 White House Conference on Children, a federally funded Children's Bureau was created, and Julia Lathrop, a prominent progressive activist, was appointed as its chieftain. The bureau disseminated childrearing literature that was in tune with the activist orientation of the new women, including Mrs. Max West's popular guide *Infant Care*. As advocates for children, progressive women, many of whom were former teachers, were also intimately involved in educational matters and increasingly pressed for school reform.[11]

Although the new women increasingly became involved in educational reform, "progressive education," like progressivism writ large, was multidimensional and full of competing and often contradictory ideas. Progressive school reform focused on both educational governance and the curriculum, although the two were intimately interwoven. At the administrative level, the drive for an organizational revolution began to take shape by the end of the nineteenth century. While common school reformers like Horace Mann and Henry Barnard sought to centralize their state systems of education during the antebellum years, this reform impulse was given new life in the Progressive era when centralized organization became almost synonymous with modern efficiency. Using the emerging corporations as an organizational model, "administrative progressives," as David Tyack has termed them, hoped to streamline ward or district-controlled schools by bringing them under the bureaucratic management of one large educational system, particularly in the urban areas of the nation. Clamoring for the centralization of urban schools, administrative progressives recognized the difficulty of managing the "one best system" through large, ward-elected boards of education. These lay school boards, managerial progressives such as James Jackson Storrow of Boston insisted, were "clumsy and antiquated" as well as corrupt. The solution, therefore, was to create smaller school boards, like corporate boards of directors, and to have an educational expert, a superintendent, efficiently manage the boards' initiatives.[12]

The idea of smaller, more "efficient" school boards partially reflected a lack of faith in the democratic process, and such sentiments encountered a fair bit of popular resentment. Ward leaders, populists of various stripes, and numerous progressive women opposed these undemocratic tendencies. In 1901, Ella Flagg Young, a teachers' advocate in Chicago, objected to the top-down reforms of administrative progressives because those at the bottom of system, teachers and students, were not being heard by the educational "aristocracy." Moreover, Young argued that "[t]he school cannot take up the question of the development of training for citizenship in a democracy" when the school itself functioned so autocratically. Despite the opposition, administrative progressives were quite successful at restructuring school boards in a number of cities throughout the United States. Some urban areas reduced the number of board members significantly and required that they be elected "at large" instead of by district, a move that virtually ensured that only prominent men, rather than ethnic leaders in particular wards, would be elected. In some cities, the mayor appointed the school board, which further removed education from the democratic process. In the early twentieth century, Boston's public school system sought to reduce its school committee to five members, members who would be appointed by the mayor. Forbidden to participate in national or state elections, women, in cities such as Boston, were allowed to cast their ballots only on educational matters, and Bostonian women's groups thwarted the reform because such a move would mean total disenfranchisement.[13]

Like all reforms in a pluralistic and democratic society, the endeavors of the administrative progressives were reworked and negotiated. In cities such as San Francisco, Chicago, and New York, where school boards were (at times) appointed rather than elected, mayors found that the doling out of appointments along religious and ethnic lines was politically expedient because an unbalanced board generated resentment, which actually undermined the progressive ideal of creating a homogeneous school committee composed of the efficient-minded elites of a city. Paralleling the reforms in the federal civil service, the new educational structure was often laid upon the older system, thus perpetuating inefficiency. In addition, one of the central aims of organizational reform was, as Storrow sought, "to take the school system out of politics," an aim that was quite impossible indeed. Political battles continued unabated, battles that focused on numerous educational issues including teachers' unions. The new bureaucratic structure could not even halt the political maneuvering within the system; it merely reformulated the nature of political debate. The organizational reforms of the administrative progressives simply fostered the forms of intra-bureaucratic politics that were common in other complex organizations.[14]

The organizational transformation of school governance was intimately linked with curricular reform during the Progressive era. Some of the leading administrative progressives, such as Stanford University's Ellwood Cubberley, were also advocates of curricular change. Moreover, educational organizations, such as the National Education Association (NEA), reached out to the nation's school boards in order to instruct them on curricular matters. In the early 1890s, the NEA's Committee of Ten, which has long been viewed as a bulwark of progressive curricular changes in the nation's high schools, was partially an attempt at providing school boards with expert knowledge regarding curricular improvements. From this perspective, the Committee of Ten was in tune with progressive thought. Nevertheless, the NEA committee, with President Charles W. Eliot of Harvard at the helm, largely recommended a traditional, "humanist" curriculum, a curriculum that sought to develop students' intellect and moral character.[15]

There were, of course, several strains of thought in the late-nineteenth and early-twentieth centuries that sought to break the hold of the traditional curriculum in America's schools, which prompted the label "progressive." As in the larger reformist movement, however, a consensus did not exist about what "progressive education" meant. Instead, there were, as historian Herbert Kliebard has noted, a few "different interest groups competing for dominance over the curriculum and, at different times, achieving some measure of control." One such interest group, the child-centered progressives, looked to science, particularly the developing field of psychology, as a guide for curricular reform. Child-centered reformers, such as the eminent psychologist G. Stanley Hall and his followers, argued that stage theories of child development should outline the contours of the educational system in America. Child-centered progressives, as the movement evolved, also came to argue that the interests of students should drive curricular concerns. Early leaders, such as Francis Parker of the Quincy, Massachusetts, schools, believed that children learned best through play and other interesting activities. Later, men like William Heard Kilpatrick, an educational philosopher at Teachers College, expanded on the notion of making the schools more engaging for students and helped develop curricular improvements like the project method for teaching science. The child-centered wing of progressive education eventually coalesced into a national organization, the Progressive Education Association (PEA). Founded in the first decades of the twentieth century, the PEA named Eliot as its first honorary president, followed by the eminent philosopher John Dewey. While Dewey's notion of education was complex and broad, he did, at times, outline ideas that fit the agenda of many child-centered progressives. For Dewey, education was ultimately "the enterprise of supplying the conditions which insure growth," and because "there is nothing to which growth is relative save more growth, there is nothing to which education is subordinate save more education." Dewey's philosophy certainly resonated with educators who sought to make schooling more in tune with the interests of students, and his Laboratory School at the University of Chicago provided inspiration. Although often perceived as a strict child-centered progressive, Dewey actually discarded the dualism between the traditional and the "new" curricular approaches. Instead of "*Either-Ors*," with "no intermediate possibilities," America's leading philosopher found shortcomings and possibilities in both forms of education.[16]

Like the child-centered group, the "social-efficiency progressives," as Kliebard has called them, also rejected the traditional, nineteenth-century curriculum. Leaders of this orientation, such as Cubberley, John Franklin Bobbitt, David Snedden, and Charles Prosser, used a corporate model of efficiency to end the wastefulness of the public school curriculum, as did the administrative progressives to reorganize educational governance. Like the child-centered progressives, social-efficiency advocates based their reforms on science, but it was not the theoretical psychology of Hall; it was an industrial "science of exact measurement and precise standards in the interest of maintaining a predictable and orderly world." Perhaps nothing epitomized the social-efficiency agenda more than the school survey. Growing in popularity at the beginning of the twentieth century, educational experts analyzed a community in order to discover its present conditions and sought to mold the schools to fulfill the occupational needs of the region. Not surprising, the school surveys often recommended that educational institutions improve their vocational preparation to satisfy the demands of local industries. A survey of the El Paso schools, for instance, suggested that girls, particularly Mexican girls, needed more in the way of domestic preparation. Bobbitt's survey of the San Antonio schools captured the social-efficiency ideal. "People," Bobbitt recommended in 1915, "should be taught at public expense to write only so well as they need to write for carrying on their various daily affairs. This means that that clerical people...should be trained to high quality and high speed[;].... factory workers, farmers,...seamstresses, laundry workers, [and] housewives...need only to write a simple plain hand with only a moderate amount of speed." For Bobbitt, curricular differentiation was the key to an efficient "[d]emocratic education," which entailed "giving everybody an equal opportunity for the education which he actually needs."[17]

Discovering what type of education students "actually" needed and thereby fitting them for their lot in life required some precise sorting mechanism. By the early twentieth century, intelligence tests came to perform that function. Lewis Terman, a psychologist at Stanford University, revised Alfred Binet's mental test and developed the concept of an intelligence quotient, a concept that determined "the ratio of mental age to chronological age." Terman and others, such as H. H. Goddard, began administering intelligence tests to individual children early in the new century, but the arrival of World War I dramatically altered the tests' uses. Army officials wanted to utilize mental tests on large groups of recruits in order to discover those unfit for service and those with the potential for leadership, and America's psychologists rose to the challenge by devising "group tests." Once the logistics of group testing was worked out, Terman noted in 1919, "little special psychological training either for giving the tests or scoring them" was needed. Therefore, group intelligence tests could be given to school children to determine what type of curriculum the students "actually" required. Terman argued that "every pupil [should] be given a mental test within the first half-year of his school life" and because, as the psychologist insisted, intelligence remained relatively constant throughout life, the test would accurately predict what variety of schooling the child ought to have.[18]

The reformist impulses of the social-efficiency progressives, as well as child-centered advocates, gained national prominence in 1918 with the publication of the Commission on the Reorganization of Secondary Education's *Cardinal Principles of Secondary Education*. The commission, organized by the National Education Association and the U.S. Bureau of Education, recommended a curriculum that did not solely focus on students' intellectual development, but rather on pupils' overall wellbeing. The *Cardinal Principles* report outlined a variety of "objectives" that American schools should emphasize in order to address the needs of the whole child, including vocational training and leisure activities. These recommendations not only intended to make education more efficient, but also more in tune with the interests of students, aims both groups of progressives sought in order to undermine the traditional, academic curriculum. Although the curricular innovations of educational reformers had captured national attention, this did not mean that students and teachers around the nation automatically accepted the new educational directions. Although vocational education increased dramatically, working-class students in Michigan, as David Angus and Jeffrey Mirel have found, often did not take the industrial training courses that many social-efficiency advocates helped create in order to prepare pupils for their stations in life. Instead, many of these students used education as a means of entering the white-collar professions. Despite the dramatic talk about educational reform, classroom practices sometimes remained quite traditional well into the twentieth century, but, nevertheless, the *aims* of schooling had changed considerably during the Progressive era.[19]

An American School

In the decades surrounding the turn of the twentieth century, enroll-ment in America's public schools exploded. Progressive reforms, such as compulsory schooling legislation, child labor laws, and better enforcement of those regulations, partially contributed to this growth. In 1909, for instance, New York's revised school law required that "[e]very child between seven and sixteen years of age...shall regularly attend...a school in which at least the six common school branches of reading, spelling, writing, arithmetic, English language and geog-raphy are taught in English." Similarly, Illinois's 1889 Edwards Act— although offensive because of its requirement of the exclusive use of English and, therefore, partially repealed—strengthened the state's compulsory education requirement and established truant officers to enforce the law. By 1890, the enrollment in Chicago's public and private schools increased by over twenty-three thousand. As impor-tant as the progressive reforms were, the growth of the student pop-ulation was primarily a result of immigration. Enrollment in New York City's schools, for example, jumped 60 percent during the fif-teen years before 1914, largely because of the population boom in the immigrant neighborhoods on the Lower East Side. The arrival of large numbers of immigrants from southern and eastern Europe par-tially facilitated the "progressive" changes in education. With public schools overflowing, new pedagogical techniques, efficient manage-ment, and more centralization were required. More than anything, the arrival of the immigrants inspired a drive for cohesion which, as many believed, could only be achieved through the standardiza-tion of curricula and, especially, an increased focus on the American language. These progressive changes naturally worked against public bilingual schooling. In an era in which a new unified "American" school was being constructed, the polyglot boardinghouse came to be seen as an outmoded institution, nearly ready for demolition. After all, as one North Dakota school official noted in 1896, the modern school sung praises to the ideal of "[o]ne country, one language, [and] one flag."[20]

The American school, of course, was not the only element that was unraveling bilingualism in the United States. During the Progressive era, the nation had become an increasingly connected society, and the conformist culture of the day—even the cities, Dreiser com-mented, appeared "so much alike"—was beginning to work its magic on many of America's foreign-language speakers. Numerous

German immigrants reported that their children were becoming Americanized, particularly with regard to language. In 1890, an immigrant in Minnesota confessed that his children "speak almost no German, but they do understand some of everyday conversation." In the years before World War I, the Scandinavians were also shedding some of their linguistic heritage and becoming more comfortable with the American language. Even the new immigrants from southern and eastern Europe were quickly becoming assimilated during the Progressive period. In a study of schools in twelve small cities (save Chicago) in the Northeast and Midwest, the U.S. Immigration Commission found that by 1911 nearly all of the pupils with foreign-born fathers spoke English. Moreover, the fathers often spoke the American tongue as well. Over 90 percent of the Scandinavian, French-Canadian, and German-Jewish fathers understood English; nearly 90 percent of the German and Dutch fathers also knew the language of their adopted country. Russian and Hungarian fathers were the least familiar with English, with only 56 and 60 percent, respectively, able to speak English, which, nevertheless, was still quite astounding. Perhaps most surprising was that English was spoken at home among nearly half of the foreign families studied. The vast majority of the Dutch, German-Jewish, German, and Scandinavian families used English in their private lives, while Russians, Poles, Slovaks, Lithuanians, and Hungarians utilized the language at home less than their "older" immigrant counterparts.[21]

Part of the reason for the rapid assimilation of newcomers during the late-nineteenth and, especially, early-twentieth centuries was the rise of a popular culture. By hanging out with their peers at school, at the nickelodeon, in the theater, and in the streets, foreign children learned American mores and wanted to shed their "greenhorn" label by adopting the majority culture quickly. Immigrant children, as one parent noted in 1908, were exposed to many "temptations" by the Americans, and perhaps none was more powerful than the modern style of dress. Addams suggested that many of the foreign children appearing in Chicago's juvenile court, most of whom were German and Polish, were there because of unlawful activities conducted in order to obtain "Americanized clothing." In 1916, Dreiser asked, "What becomes of all the young Poles, Czechs, Croatians, Serbians, etc., who are going to destroy us?" His humorous answer was that "[t]hey gather on the street corners when their parents will permit them, arrayed in yellow or red ties, yellow shoes, dinky fedoras or beribboned straw hats." These foreign children were no threat because "[w]hatever their original intentions may [have] be[en], they can't resist the American yellow shoe, the American moving picture, 'Stein-Koop' clothes,...the popular song, [and] the automobile." "Instead of throwing bombs or lowering our social level, all bogies of the sociologist," Dreiser concluded, immigrant youngsters "would rather...smoke cigarettes, wear high white collars and braided yellow vests and yearn over the girls who know exactly how to handle them." This acceptance of popular culture meant that the immigrants were, in fact, "fast becoming Americans."[22]

Popular culture was not the only mechanism of assimilation, however. During the Progressive era, many public schools in the United States diligently embarked on a campaign to Americanize foreign-language-speaking students. The progressive reforms that called for the centralization of schools and the standardization of curricula were, at their heart, making uniformity and cohesion the cornerstones of U.S. education. Alternatives to the "one best system," as Tyack has called the reformers' vision of homogeneity, were perceived as outdated and inefficient. In a society that worshipped all things modern, bilingual education increasingly came to be seen as an artifact of a bygone era, and many schools throughout the nation revised their language policies. These progressive educational reforms, driven by the society's obsession with order, were, in many respects, facilitated by the arrival of the new immigrants, but invariably impacted other foreign-language speakers. Louisville, St. Louis, and St. Paul, for instance, ended their elementary German-language programs, and New York, Chicago, and Cleveland forced foreign languages out of the primary grades. Some states, such as North Dakota, developed English-only curricula to facilitate the assimilation of immigrants, as did some heavily Mexican-American counties in Texas. In 1893, the Lone Star State began to pass a series of school laws that undermined bilingual instruction, but the English-only legislation was ignored in many areas, as was the Americanizing curriculum in North Dakota. Yet, the emphasis on English was not only an attempt at assimilating foreign-language speakers; the emphasis itself was seen as a progressive reform because English, in an increasingly impersonal America, became one of the central skills students needed in order to navigate their way through modern society. In 1896, California's Committee on the Course of Study for the Elementary Schools noted that students "should be given command of the tools on which future progress depends," and, for the committee, the primary tool was a knowledge of English.[23]

As English came to be seen as an urgent necessity for modern America, cities with large populations of southern and eastern European immigrants developed new educational programs to give foreign-language-speaking students the gift of the American language as quickly as possible. During the decades surrounding the turn of the twentieth century, Cleveland's population of southern and eastern Europeans increased by over 600 percent; most of the newcomers were Bohemians, Italians, Hungarians, and Poles. To address the special linguistic needs of the new students, Cleveland's public schools established "steamer" classes. In 1901, the principal at the Harmon School began the city's first steamer class, a class so named because the students had just arrived in the United States via steam ships. The classes, which were intended to quickly transition students to English, grew in number during the first decades of the century; by 1915, there were about twenty-five. Herbert Adolphus Miller, a professor of sociology at Oberlin College, noted in 1916 that Cleveland's schools needed to "increase the efficiency" of these steamer classes, which, according to Miller, was best demonstrated by the work of New York City's public schools.[24]

Like Cleveland, the public schools in New York began special classes for the new immigrants at the beginning of the twentieth century. Greatly facilitated by the work of Julia Richman, who was appointed as a district superintendent for the Lower East Side in 1903, the steamer classes—or "C classes" as they were more often called in Manhattan—brought together foreign-language-speaking students in order to quickly teach them enough English to transfer to regular classrooms. Students in the C classes were only allowed to remain in these special English-only classes for an academic year or less; if they failed to acquire the linguistic skills needed they were ushered into D classes. Many students, however, became fluent enough in the American language to enter regular classes after about six weeks. The English-only aspect of the New York schools was rigidly enforced. Some students caught speaking Yiddish, for instance, had to have their mouths washed out with kosher soap. By 1913, Manhattan had over sixty C classes, and Brooklyn had nearly twenty. The impetus for the C classes came not only from New York's school establishment but also from the Americanized Jewish community in the city, particularly those of German ancestry. The Educational Alliance, a Jewish philanthropic organization, had begun private Americanization classes in 1890 in order to prepare students for the English-only public schools. Although partially driven by altruistic motives, the

Educational Alliance—with which Richman was involved—and other philanthropic institutions also pushed for the Americanization of the eastern-European Jews in order to salvage their status in the United States; backward Jewish peasants, some prominent Jews worried, might reflect badly upon those who had already assimilated and become successful in their adopted country. The C classes, therefore, served a twofold purpose. They provided students with the linguistic skills that were increasingly needed in modern America, and they attempted to preserve the social position of the established Jews by cutting new students off from their potentially embarrassing traditions and languages.[25]

While the move away from bilingual instruction was a reflection of both the nation's changing demographics and the perceived skills needed for success in modern America, the Progressive era was also a period in which science was elevated to a godly status, and the emphasis on English, at least for the foreign-language speakers, had a scientific basis. When Miller criticized Cleveland's steamer classes for being less efficient than those in New York, he was primarily pointing to the lack of scientific grounding for Cleveland's program. "The educational officers of the city," Miller noted, "have never worked out any special methods for teaching English to these non-English-speaking children." By the late-nineteenth century, new methods were supplanting the older focus on translation and grammar. Promoted in the United States by the philologists in the Modern Language Association (MLA), the "natural" or "direct" method emerged as the standard and "scientific" means of teaching a second language. Teaching a foreign language through the natural method entailed immersing the student in the language so the pupil could make direct associations between actions and the linguistic meaning, rather than translating in the mind. The science of the day also suggested that young children had the "plasticity" to acquire a language naturally, which implied that language study should begin as early as possible. Charles F. Kroeh, a prominent language scholar at the Stevens Institute of Technology, noted in the late-nineteenth century that the natural "method consists in speaking only the foreign language in the class room, as though English were not in existence."[26]

While Kroeh was primarily concerned about the teaching of foreign languages to English speakers, the same methods were applied to foreign-language speakers learning English. Addressing the Modern Language Association in 1890, President Eliot of Harvard insisted that "[o]ne of the chief objects of your association has been the promotion of English." Eliot was quite right; the research of the MLA and other linguistic scholars was increasingly being used to bolster the American language. English-only classes for foreign-language speakers emerged from the science of linguists. Drawing on the scholarship of MLA researchers and their European counterparts, particularly the French scholar François Gouin, American advocates of steamer-type classes applied the natural method to America's immigrant students. One of Gouin's American followers, Frank V. Thompson, argued that the French linguist's methods were based on "sound pedagogical principles" and, therefore, could be introduced into America's multilingual classrooms. Stressing the spoken language and deemphasizing phonics, translation, and grammar was the key. Thompson noted that "in teaching a Pole the meaning of the word 'door,' a teacher who uses the direct method leads the pupil to associate the word 'door' directly with the object 'door,' suppressing as far as possible connection with the word for 'door' in Polish." This, of course, meant that "teaching English to pupils" required that "only English" was permitted in the classroom. By and large, these were the methods of New York's C classes, unlike the hodgepodge pedagogy in Cleveland's steamer courses. Although there was still some phonics in the C classes, the focus, as New York's superintendents made clear, was on object lessons, and many future teachers learned the direct method through their student teaching experiences.[27]

A Polyglot School

Progressivism was certainly chipping away at the foundations of the polyglot boardinghouse; localism was gradually eroding, and the American language was becoming the central focus of modern public schools. Moreover, many of the new foreign-language speakers in the United States did not see education—or bilingual education for that matter—as their main priority. Numerous southern Italian families, for example, were leery of state-sponsored schooling because of their experiences in the Old World and, instead, opted to send their children to work. Besides, many newcomers hoped to return to Italy after they had saved enough money; working children—as opposed to idle children sitting in school—facilitated this goal. Many Slavic youngsters also entered the labor force instead of the school, often because home ownership, at least for the first generation, was seen as the most pressing concern in the new environment. The patriarch Jurgis, Sinclair's central character in *The Jungle*, epitomized this pattern. He and his family saw buying a home as the ultimate form of financial security in America, and children's earnings, at least after the family fell on hard times, could help with the payments. Although there were child labor regulations, Sinclair wrote, the "law made no difference, except that it forced people to lie about the ages of their children."[28]

In spite of all of this, bilingual schooling persisted. Private educational institutions, which had long been arenas for language maintenance, fostered the languages of many of the new and old immigrants. In Cleveland, for instance, over seventeen thousand pupils maintained their mother tongues through language instruction in the city's Catholic schools in 1915; Jewish students and other non-Catholics also had private schools that cultivated their non-English languages. In Hawaii—a territory of the United States since 1900—numerous private language schools served the linguistic needs of the large Japanese population, and a smaller number of schools maintained the languages of Korean and Chinese immigrants on the islands. These language schools, often in operation before and after public school hours, educated tens of thousands of pupils bilingually by the early decades of the twentieth century. Private schooling was also common in Louisiana, which sustained the French language to such a high degree that, in some parts of the state, Anglos actually assimilated to the Creole and Acadian cultures, rather than vice versa. In the public realm, bilingual schooling also continued. In isolated ethnic communities, dual-language instruction persisted in spite of English-only laws and state-mandated curricula. In Texas, Nebraska, North Dakota, and throughout the Midwest, local schools offered bilingual education even though state and county officials frowned upon either any foreign-language instruction or, as in Wisconsin, extensive bilingualism. In New York's English-only regular classrooms, student-led dual-language instruction frequently occurred. Children with no knowledge of the American tongue, as one observer noted in 1906, were "seated next to compatriots, who act[ed] as interpreters for them."[29]

While enclaves of foreign-language speakers continued with their tradition of bilingualism, areas that had a strong ethnic intelligentsia resisted the English-only trend more systemically. In New Mexico—where *hispanidad* leaders actively promoted the Spanish language and culture—dual-language instruction successfully endured, even in the face of adversity. As the territory sought statehood in the early-twentieth century, Anglos came to dominate the higher levels of educational administration and pushed for the sort of progressive reforms that were then en vogue, including a standardized English-only curriculum. These reforms, Anglo leaders in New Mexico believed, would alleviate some of the concerns congressmen (particularly Indiana's Senator Alfred Beveridge) had about New Mexico's adherence to the Spanish language. By 1907, Superintendent James E. Clark encouraged the adoption of a uniform series of English textbooks and began the process of consolidating the rural schools; the aim, of course, was "making the public school an all-English school." These reforms, the superintendent noted, better "equipped" students for life in modern America, its "business, social, [and] political" dimensions.[30]

In spite of the drive for an English-only curriculum, the powerful Spanish-American leaders of the territory ensured that bilingualism was an enduring feature of New Mexico. Becoming part of the United States in 1912 did not hamper educational documents from being published in Spanish, and the new state constitution allowed for bilingual instruction in the public elementary schools. A Spanish-English curriculum, therefore, was a regular aspect of the state's common schools in heavily Spanish-speaking areas. In Guadalupe County, for instance, bilingual education continued unabated, although, as one school official noted in 1910, educators were "using the Spanish language a little too much in teaching." Taos County was the epicenter of the state's public bilingual education efforts. The teachers there were "specially adapted for teaching Spanish-American children," because many of them were native Spanish speakers themselves. Moreover, the county was home to the Spanish-American Normal School located in El Rito (now in Rio Arriba County). Established in 1909, the normal school educated future teachers for the demands of bilingual instruction and primarily enrolled "Spanish-speaking young men and women" for these endeavors.[31] In an impersonal society where credentials and expertise were a necessity, the Spanish-American Normal School was in tune with progressive ideals; it fostered the educational science and professionalism that were at a premium, but it simultaneously bolstered Spanish-speaking leaders' agenda: bilingual education.

Other ethnic groups in the United States used a similar strategy for resisting the English-only movement. Liberal German leaders in particular were adept at transforming progressive language and science to retain their dual-language programs. By the turn of the twentieth century, public German-language instruction was most pronounced in the towns and cities that had been traditional Forty-eighter strongholds, such as New Braunfels, Texas, Belleville, Illinois, New Ulm, Minnesota, and cities throughout Indiana and Ohio. Even Cleveland, with its English-only steamer classes for the "new" immigrants, continued elementary German studies. The state laws in the Midwest that allowed for the German language in the public schools certainly facilitated bilingual instruction, but it was primarily the maneuvering of local ethnic leaders that made dual-language education relevant to the new "progressive" times. Cincinnati, with its model bilingual program, increasingly made use of the direct method—the method that formed the scientific basis for English-only classrooms—for the teaching of German. Ethnic leaders, such as Supervisor of German Heinrich Fick, understood the progressive educational trends of the day and utilized them as the grounding for German-English instruction. The German "language," Fick noted in 1911, "is taught as a living tongue, all the communication between teacher and pupils being conducted in the language to be acquired." Fick also emphasized the bilingual educators' professionalism and pedagogical expertise. "Teachers conversant with this [direct] method," Fick stated "are trained in the National German American Teachers' Seminary in Milwaukee, Wis., and also in the Teachers' College of this city." With the German-English instruction grounded firmly in progressive educational theory, the program flourished during the decades surrounding the twentieth century. Until World War I, Cincinnati continued to annually educate thousands of students bilingually.[32]

The ethnic leaders of Indianapolis's German program were even more explicit in the usage of progressive ideals for the justification of bilingual education in the city. In the early twentieth century, the supervisor of German, Robert Nix—from the Forty-eighter town of New Ulm—understood the educational mood of the nation and worked tirelessly to keep German-English instruction germane to Indianapolis's public-school curriculum. As elsewhere in the country, the focus on the English language and other progressive reforms made its way to the Hoosier capital in the decades surrounding the turn of the twentieth century. Nix, however, tried to keep bilingual instruction in tune with the changes. The supervisor of German was adamant about the professional development of his teachers, and numerous seminars, meetings, and examinations ensured that the German educators were experts in their field. Moreover, the new linguistic methods came to hold a central place in the pedagogical practices of the program. The "'direct' method," Nix noted in 1909, "endeavors to equip the pupil with a practical vocabulary, to develop in him what the Germans call 'Sprachgefuehl' [a feel for the language]," and this was the method of choice in Indianapolis. To justify elementary bilingual instruction more forcefully, Nix drew upon the scientific findings of the scholars affiliated with the Modern Language Association. "In childhood," Nix quoted from the MLA's Committee of Twelve, "the organs of speech are still in a plastic condition." Learning a language, therefore, required study at an early age, because any "later period of youth is distinctly a bad time to begin." With the German program aligned with progressive thought, the enrollment grew dramatically until the United States entered World War I. In 1909, just over 6,500 elementary pupils received German-English instruction, and by 1916 nearly 8,200 students were educated bilingually.[33]

This strategy of using the language of progressive education to bolster German instruction was echoed by many in the nation. When the elementary German-language program in Cleveland's public schools was attacked in 1905 and 1914, the German community, like that in Indianapolis, used scientific and utilitarian justifications to defend the language. Cleveland's German Americans noted that science proved that language study should begin at an early age, a knowledge of German helped students understand English more fully, and the language was essential for modern commerce. Although armed with these progressive-oriented arguments, the beginning period for city's German program was pushed back from the primary years to the fourth grade. M. D. Learned, an Anglo professor of modern languages at the University of Pennsylvania, put forth many of the same arguments as the Cleveland Germans. He recognized the need for English in modern America but insisted that German, for a student, "will enormously facilitate his study of English." Moreover, Learned argued that German had become "the language of international culture," which made it a necessity for business and scholarship. A knowledge of the German language, the professor suggested, "would increase the efficiency and prosperity of our great country to a degree of which we have never yet dreamed." During the Progressive era, German-English instruction not only required a new justification, it also demanded caution. In an 1898 pamphlet published in Germany, Karl Knortz, the supervisor of German in Evansville, Indiana, noted that in the English-obsessed climate "discreet teachers" were needed for maintaining the mother tongue in America. This sentiment was echoed by the National German-American Teachers' Association. The association, which had long been an outspoken and unabashed champion of linguistic maintenance, increasingly tempered its language and emphasized the necessity of English, as well as German, for life in industrial America.[34]

Discretion was not the tactic for the ethnic leaders in Milwaukee, however. As in Cincinnati and Indianapolis, school officials in Milwaukee made use of the language of progressivism to justify foreign-language instruction, but they drew on an entirely different strand of progressive thought. Instead of the efficiency and scientific language that was transformed to bolster bilingual education in other areas of the nation, Milwaukee's ethnic leaders emphasized the child-centered and humanitarian side of progressivism to validate their elementary foreign-language programs. To a large degree, this emphasis resonated because leftist reform was taking root in the city, particularly under the guidance of the powerful Social Democratic Party. The Milwaukee of the early-twentieth century was a community where Socialists such as Meta Berger—the wife of the famed Victor Berger—sat on the school board. Leftist activists forged relationships with other reformers in the city and reached out to ethnic organizations such as the Polish School Society. By the 1910s, Milwaukee's elementary public schools offered not only German, but Polish and Italian as well. Unhappy with parochial instruction, the city's ethnic Socialists were instrumental in the pushing for the adoption of Polish in the elementary curriculum. The introduction of the languages of the "new" immigrants was also the sort of progressive reform for which child-centered leaders like Jane Addams called. Early in the century, Addams argued that schools should do more to preserve the traditions of the newcomers, partially because to do so ensured that family relations were not severed by harshly implemented Americanization efforts, but also because maintaining linguistic and cultural traditions was a genuinely humane endeavor. In Milwaukee, the inter-ethnic cooperation of leftist school reformers made its mark. In 1916, the city employed one hundred twenty-six elementary German teachers, sixteen elementary Polish teachers, and four elementary Italian teachers; together, these educators gave foreign-language instruction to over thirty-five thousand pupils.[35] These updated justifications and the keeping of bilingual education in line with progressive thought preserved dual-language instruction, and in many localities throughout the nation it flourished. World War I, however, would damage the polyglot boardinghouse to a significant degree.

7

"If You Can't Fight over There, Fight over Here": World War I and the Partial Destruction of the Boardinghouse

Certain aspects of progressive thought certainly undermined public bilingual education, but nothing damaged the polyglot boardinghouse more than World War I. As Europe went to war in the summer of 1914, the United States instinctively remained neutral. There was not yet any reason to discard the nation's traditional isolationist stance regarding international affairs. A series of events, however, broke down America's commitment to isolationism and ultimately drew the country into the Great War. In 1915, a German U-boat sank the *Lusitania*, a vessel that carried more than a hundred American citizens. Although there was a public outcry for a declaration of war, President Woodrow Wilson was not ready to take that catastrophic step. By 1916, the submarine warfare had waned, and Wilson successfully ran for reelection by trumpeting his ability to keep the United States out of the European conflict. The war, however, was taking its toll on Germany, and drastic measures were needed. The German military and political leaders announced on January 31, 1917 that unrestricted submarine warfare would resume in an effort to cut Great Britain's supply lines. Americans were outraged, and Wilson quickly severed diplomatic ties with Germany. Throughout the nation, newspapers praised Wilson's decision. "The President," St. Louis's *Globe Democrat* declared, "has done the only thing that in honor could be done....and has spoken with the voice of the American people." The *Philadelphia Record* stated that "[t]he American people stand solidly with the President in what he has done, and in what he may have to do."[1]

Waiting for what the president had to do did not take long, only two months in fact. These two months—February and March—witnessed developments that brought the American people out of their isolationism. In January, the German foreign minister, Arthur Zimmerman, sent a secret telegram that suggested Mexico and Japan should join the conflict on Germany's side. The telegram, which was intercepted and then leaked to the public in early March, stated that "we shall make war together and together make peace. We shall give general financial support, and it is understood that Mexico is to reconquer the lost territory in New Mexico, Texas, and Arizona." The Zimmerman note coupled with the sinking of four U.S. ships by U-boats in February and March pushed America and its president too far. On April 2, Wilson called for war, noting on Capital Hill that "I advise that the Congress declare the recent course of the Imperial German Government to be in fact nothing less than war against the government and people of the United States; that it formally accept the status of belligerent which has thus been thrust upon it." Victory, Wilson famously stated, would help make the world "safe for democracy."[2]

The declaration of war unleashed a zealous form of patriotism that not only threatened democracy at home, but that also made the nation unsafe for some American citizens. George M. Cohan's popular wartime song "Over There" encouraged "Johnnie to get your gun,... [and] show the Hun, you're a son-of-a-gun," but Johnnie did not only fight "over there." "If you can't fight over there," the war slogan told Americans, "fight over here." And fight they did, against all aspects of American society that were deemed "un-American," particularly against the foreigners and their languages. In 1917, Theodore Roosevelt noted, "We call upon all loyal and unadulterated Americans to man the trenches against the enemy within our gates." Educators took Roosevelt's message to heart. Addressing the Indiana State Teachers Association, Professor James Woodburn pleaded "to save America from being a polyglot nation—a conglomeration of tongues and nationalities, like a 'polyglot boardinghouse,' as Mr. Roosevelt has put it." In the trenches at home, Americans were encouraged to destroy the boardinghouse. "We must have but one flag," Roosevelt noted. "We must also have but one language. That must be the language of the Declaration of Independence."[3]

Propaganda, Patriotism, and Paranoia

Although a declaration of war was no small matter, going to war would require a tremendous effort. The United States, in its relatively brief history, had never participated in a modern, total war. The world, in fact, had not yet seen such a war until 1914, the sort of war that required a belligerent nation to devote all its energies toward the conflict. Prior to the outbreak of the Great War, the progressive reformers in the United States had been working for decades to create a more centralized and efficient nation, but their work had not been completed by the time of the American involvement in the international struggle. The country, therefore, still partially lacked a centralized structure that could quickly mobilize the United States for war. The days of America as the land of "island communities" were not long past, and localism in the early-twentieth century certainly persisted. But, a new type of war was upon the nation, a war of attrition, machine guns, trenches, barbed wire, and poison gas; a war that, as John Dos Passos wrote, was not "fun anymore," a war that was truly "grim." The French writer Louis-Ferdinand Céline also stated that the war was "no fun anymore," particularly as his protagonist realized that the "collective murder" and the "infernal lunacy could go on for ever." In order to gear the country up for this new war, historian David M. Kennedy has persuasively argued, the Wilson administration, lacking a centralized structure, drew upon what was at hand, and what the government had at its fingertips was the progressives' weapon of choice: publicity. The muckraking journalists had long been writing exposés to drum up popular support for particular issues, and now the president, whose support largely came from the progressives, made use of the same techniques to sell the war and his brand of "patriotism" to the people.[4]

Propaganda became one of the central mechanisms for mobilizing the American public for war. The British propaganda machine was a powerful force in the United States. As the war dragged on, the lines of communication with the Central Powers were disrupted, giving the Allies a monopoly over war news and helped win the Americans over to their side. As the United States geared up for war, the nation developed its own propaganda machine. For the purpose of molding the American mind, President Wilson established a propaganda bureau, the Committee on Public Information, and selected George Creel, a muckraking journalist and loyal Wilsonian progressive, to head the new organization. "[T]he war," Dos Passos sardonically wrote, was "fought from the swivel chairs of Mr. Creel's bureau in Washington." Through a plethora of reports and pamphlets, Creel's committee, with the secretaries of state, war, and navy as members, generated a great deal of support for the war effort. One of the Committee on Public Information's central goals was to create a united and cohesive nation, long an aim of progressive reformers such as Creel. Unity was largely achieved by cultivating patriotic sentiment, and the American public took the bait, although it certainly did not need much of a push. The "four-minute men," Creel's public speakers who drummed up support for the war, helped bring the message of patriotism and unity to the masses. Americans not only became patriots after war was declared, they became superpatriots and whole-heartedly—almost religiously—devoted themselves to wartime endeavors.[5]

Once the American propaganda machine stirred up hysterical patriotic feelings, those emotions took on a life of their own and manifested themselves in extreme forms. Local patriotic societies popped up all over the nation, and the American Protective League rooted out disloyalty with the blessing of Attorney General Thomas W. Gregory. Most suspected of divided loyalties, as Wilson noted in 1915, were the "hyphenated Americans," particularly the German Americans, whom the novelist Owen Wister called the "Kaiser's helpful hyphens." Wilson toned down his rhetoric regarding "hyphens" for the 1916 election, and his declaration of war underscored that a "friendship" remained with "the millions of men and women of German birth and native sympathy who live amongst us and share our life." But, the president added, "[i]f there should be disloyalty, it will be dealt with with a firm hand of stern repression." Reflecting on the war years, immigrant Ludwig Dilger blamed British propaganda for the promotion "of hatred of everything German." So fanatic was the hatred that even German words became the object of abhorrence; "sauerkraut," in this zealous atmosphere, was deemed "liberty cabbage." The property of German Americans was vandalized, and an effigy of the Kaiser, in a rural area of North Dakota, was hung, buried, and, then, "dynamited," which brought forth the participants' response: "That's the last of [the] Kaiser." After the *Lusitania* was sunk by a German submarine, a German pastor was murdered by fanatical Anglos in Indiana because of his pro-German sympathies. In April 1918, a nativist mob lynched a German immigrant, Robert Prager, in neighboring Illinois. Although the leaders of the mob were brought to trial for murder, the jury found them not guilty. The celebrated leftist intellectual Noam Chomsky has poignantly noted that "the Creel Commission...succeeded, within six months, in turning a pacifist population into a hysterical, war-mongering population which wanted to destroy everything German, tear the Germans limb from limb, go to war and save the world."[6]

In addition to a war-mongering patriotism, the hatred of everything German was a reflection of paranoia. Americans came to believe that the German government had devised plots against the United States. In November 1917, for instance, Professor Woodburn of Indiana University noted that the "German rulers and their hired agents in this country...have deliberately planned the invasion and partition of our territory...[and] have filled our lands with spies." These beliefs, of course, were not entirely unfounded. The Zimmerman note was authentic, and a German military attaché, Franz von Papen, encouraged sabotage within the United States. From the American perspective, moreover, Germany was a militaristic and autocratic nation, fully devoted to the war effort. The invasion of Belgium and the unrestricted submarine warfare violated the international rights of neutral nations and demonstrated the German government's willingness to win the war by whatever means necessary; for Americans, Kaiser Wilhelm's sentiments smacked of fanatical militarism, sentiments that many German soldiers took to heart.[7]

Although some of the wartime paranoia was justified, much of the fear was chimerical and fueled tales of outlandish intrigues. Rumors began to spread that the Germans in the United States were involved in plans to poison food and water supplies and to bomb the nation's infrastructure, particularly bridges and factories. Wister, along with Gustavus Ohlinger—a leading publicist of supposed German plots—noted that the hyphens "have been virtually waging war in this country." The war against America, according to Wister and Ohlinger, was primarily the undertaking of the recently arrived "Prussianized" German immigrants, those from the northern sections of Germany who emigrated for financial gain, not for religious or political freedom. Because of Germany's Delbrück Law, which in theory enabled some emigrants to retain their German citizenship, German Americans were thought to be loyal to the fatherland and subject to the commands of the Kaiser. Although the dual-citizenship law inspired concern about the loyalty of German Americans, that concern was largely unwarranted. Retaining German citizenship was not automatic; it required an emigrant to apply to the German government for the special status. Moreover, the United States' naturalization process terminated citizenship to other nations. Although numerous paranoid stories about those devilish Huns in America circulated during the war years, most of these machinations proved to be false. Even Secretary of State Robert Lansing admitted that the anxiety about German-American intrigues was baseless. Although, in general, Germans in the United States were loyal to their adopted nation, they remained under suspicion and were sometimes restricted from areas that were seen as possible targets, such as military bases.[8]

The American propaganda machine certainly whipped up support for the war, a form of patriotic support that was nearly fanatical, but additional measures were needed in order to achieve the "100 percent Americanism" that the government thought was necessary for social cohesion in a time of crisis. A national emergency called for extreme measures, and during this American crisis, as Dos Passos noted, "Wilson became the state." Congress bestowed upon the Wilson administration extraordinary powers to build consensus for the war effort, and, for the sake of national unity, dissent needed to be suppressed. True to the words in his declaration of war, Wilson ensured that those who opposed the war felt the "firm hand of stern repression." The leftist labor organization, the Industrial Workers of the World (IWW), suffered greatly from that firm hand, and, in 1917, many Wobblies, as the IWW members were called, were jailed. The Espionage Act, which was enacted in June 1917, was revised and strengthened in May 1918 and made it a crime to speak ill of the United States or the war. The Sedition Act, as the revisions to the original law came to be called, gave Attorney General Gregory even more power to disrupt the activities of those who opposed the war, particularly pacifists and socialists. Perhaps the most notorious use of the Espionage Act was the jailing of the socialist Eugene V. Debs for speaking out against the war in Ohio. Debs, who ran against Wilson in the 1912 presidential election, willingly admitted his opposition to the war at the trial and stated that he would do so even if he was its only opponent. After being found guilty, the socialist leader remained unrepentant and principled. "[W]hile there is a lower class," Debs told the court, "I am in it; while there is a criminal element, I am of it; and while there is a soul in prison, I am not free." Writing about the war years in his novel *1919*, Dos Passos wryly commented, "If you objected to making the world safe for cost plus democracy you went to jail with Debs."[9]

Wilson's firm hand of stern repression also pushed the press—particularly foreign-language publications—and ethnic organizations against the wall. Postmaster General Albert Sidney Burleson vigorously used the Espionage Act to censor subversive materials being sent through the mail. The ethnic press was monitored carefully for seditious statements, and a special division of Creel's Committee on

Public Information employed foreign-language-speaking journalists to instill patriotic sentiment among the immigrants. In October 1917, however, the Trading-with-the-Enemy Act was enacted and required the foreign-language press to provide the postmaster with English versions of its political news, which put a great financial strain on many of the smaller newspapers. Understanding the mood of nation, many publications began to censor themselves, including English-language periodicals. For the most part, the foreign-language press supported the war after the United States joined the international conflict and tried to demonstrate its loyalty by encouraging the readers to buy liberty bonds. Nevertheless, many foreign-language newspapers, particularly those written in German, came under attack and, ultimately, did not survive the war. On May 31, 1918, for instance, Indianapolis's *Telegraph and Tribüne* told its readers that it would soon cease publication because "a pronounced prejudice has arisen in this country against everything printed or written in the German language, regardless of the fact that the German language newspapers are the means of reaching thousands of persons who are reached in no other way." Like the foreign-language press, ethnic organizations, particularly the national, state, and local chapters of the German-American Alliance, came under attack. Devoted to the preservation of German language and culture in America, the National German-American Alliance, supported primarily by liberal club Germans, was accused of un-American activities, and in July 1918 Congress, with Ohlinger testifying against the organization, withdrew the association's charter.[10]

Americanizing the Schools

As a reflection of the larger society, the nation's public schools were also inundated with propaganda, patriotism, and, ultimately, paranoia. Fighting the war "over here," numerous agencies, both public and private, supplied propaganda to educators and students with the hope of "spreading the teachings of true Americanism in the war crisis." The National Education Association, the National Security League, and the National Committee of Patriotic Societies prepared educational materials to aid in the war effort. The National Security League, for instance, published a series of pamphlets to promote patriotism, leaflets aimed at teachers, and a collection of sources—*Out of Their Own Mouths*—that aimed at inspiring anti-German feelings. The Council of National Defense helped distribute materials for

Creel's Committee on Public Information, which published hundreds of books and pamphlets for "educational" purposes. With titles such as *Conquest and Kultur, German War Practices, German Militarism and Its German Critics*, and *The Prussian System*, it was clear that Creel used his propaganda machine to manufacture ill feelings toward America's enemy, as well as to inspire patriotism. In addition to the published propaganda, students themselves became proselytizers. Pupils in the Junior-Four-Minute-Men program whipped up support for the war effort just as their elder counterparts did. Even the public schools of Belleville, Illinois—once guided by the liberal sentiments of the German community, a community that questioned fanatical patriotism and condemned war as "an evil to all mankind"—organized junior speakers who "convinced their audiences of the need for . . . sacrifice and yet greater sacrifice."[11]

The National Board for Historical Service (NBHS), an organization composed of scholars and school leaders who worked closely with the Committee on Public Information, produced a wartime curriculum for elementary students in 1918. Distributed by the Bureau of Education, the NBHS's *Outline of an Emergency Course of Instruction on the War* was prepared by Charles A. Coulomb, Armand J. Gerson— both of whom were district superintendents in Philadelphia's public schools—and Albert E. McKinley, the editor of the NBHS publication *History Teacher's Magazine*. The National Board for Historical Service's *Emergency Course of Instruction* aimed at giving educators practical guidance on what to teach young children about the war and at "inculcating an admiration for the virtues of patriotism, heroism, and sacrifice." While the international conflict should be explored during lessons in the traditional school disciplines, the NBHS noted, two weekly sessions fully devoted to war issues were also necessary. For the first two elementary grades, students were to be taught the reasons for the war. "[F]ather, brother, uncle, [and] cousin . . . had to go to war," the *Emergency Course of Instruction* stated, "[t]o protect the people of France and Belgium from the Germans, who were burning their homes and killing the people, even women and children." The students' relatives also had to go overseas "to keep the German soldiers from coming to our country and treating us the same way." Older elementary students received similar justifications for the war, but also lessons on German militarism and "the work of German plotters and spies against this country." Teachers were expected to tell students that England and France were America's historic allies. The French, of course, had supported the United States during the Revolutionary

War, and Great Britain was the birthplace of America's culture and political institutions. Ohlinger paralleled these sentiments when he wrote that "England and France" were America's "parent countries." Although the United States had fought England in the Revolutionary War, the British aggression, according the NBHS, "was the work of George III and his party, not of the English people." Likewise, the historians in the Indiana State Teachers Association noted in 1917 that "[t]he colonies had originally risen against George III, a German, because of his autocratic mood."[12]

While some of the wartime propaganda for the schools stressed the villainous activities of the German Huns, it also sought to generate patriotism among teachers and students, a goal that was not hard to achieve during a national crisis. "With admirable loyalty and patriotism," the U.S Bureau of Education noted in 1918, "the schools stand ready and eager to do their full duty, whatever that may be." To ensure their patriotism, schools around the nation required teachers to take loyalty oaths, pledging their absolute fidelity to the United States. Students demonstrated their patriotism by supporting the war effort in a variety of ways. Early in the war, numerous high schools had programs to train students for military service, although pupils were discouraged from enlisting until after they graduated. Public schools asked students of all ages to help conserve materials and resources. Elementary pupils, for instance, received lessons about not wasting food and fuel and were told to "[b]e careful of [their] health" because "[d]octors and nurses are needed just now for more important work than curing children's ailments that are the result of carelessness." Students and their parents also purchased war bonds, and youngsters were encouraged to "[s]ave pennies for thrift stamps," which they did and raised vast sums of money for the war. There were less peaceful manifestations of patriotism in the schools, however. In Sylvia Plath's fictionalized autobiography, *The Bell Jar*, the narrator's "mother spoke German during her childhood in America and was stoned for it during the First World War by the children at school."[13]

The schools also demonstrated their patriotism by stepping up their Americanization efforts. During the war, Americanizing the immigrants took on a new urgency and became a national priority, one that was intimately connected with the "fight over here." The "Americanization movement," as it became called during the war years, sought not only to make foreigners supporters of an Allied victory, but also to curb the perceived Communist threat, a threat that grew exponentially with the Bolshevik takeover of Russia in

1917. Numerous public and private agencies, at all levels of society, worked to create loyal citizens of America's foreigners, making—as a skeptic of the movement noted—"Americanization...the favorite pastime of America to-day." Leading industries, such as the Ford Motor Company, were among the most zealous private organizations involved in assimilating the immigrants, but sometimes, as the critic Edward Hale Bierstadt argued, these efforts were "merely a cloak, a camouflage, which serves to hide another and far less disinterested motive." What was often hidden was industrial leaders' desire for non-threatening, loyal employees, a goal that became particularly acute after Attorney General A. Mitchell Palmer's infamous hunt for radicals in the United States. "Mr. Palmer," Bierstadt wrote in the early 1920s, "was selfish, near-sighted, frightened, and misled.... [and] thought that if he could throw the nation into a panic, making it believe that every immigrant was a potential Bolshevist and bomb-thrower, he could...pose as the saviour of a free people." Sinclair Lewis too asserted that assimilation was for the benefit of big business. The "Americanization Movement," Lewis sarcastically wrote, was designed "so that newly arrived foreigners might learn that the true-blue and one hundred per cent. American way of settling labor-troubles was for workmen to trust and love their employers."[14]

Although some had ulterior motives regarding labor struggles, many of the Americanization agencies primarily sought to teach immigrants and their children English and civics. During the neutrality period, organizations such as the Committee for Immigrants in America and the National Americanization Committee were in the forefront of the movement to assimilate the foreigners in the United States. With well-known figures such as Joseph Mayper and Frances Kellor affiliated with them, these semi-private committees increasingly assisted governmental departments during the war years when assimilation became a national obsession. Mayper, for example, conducted an Americanization survey for the Committee on Public Information in 1918. In addition, staff from the National Americanization Committee began working for the Bureau of Education in order to disseminate educational materials. Like other federal agencies, the U.S. Bureau of Naturalization also engaged in a great deal of Americanization work. Howard C. Hill of the University of Chicago concluded in his 1919 study of Americanization education that most federal, state, and local programs stressed "the acquisition of the English language and American citizenship, and...the adoption of American customs, standards, and methods of life." Hill, however, criticized the movement because the work of the various agencies overlapped considerably and did not even come close to reaching all of the immigrants. Moreover, "[s]pecial instruction," Hill recommended, "should be provided in normal schools, colleges, and universities to fit teachers for the work of Americanization."[15]

As in the larger society, patriotism in the schools, particularly the drive for Americanization, spiraled out of control and became a fanatical effort to end all elements deemed foreign. Some of this fanaticism was driven by the social paranoia during the war years that suggested that conspiracies, even conspiracies in education, were underway. The fear was that foreign thought had infiltrated the American school. Particularly harmful was that German culture—which, according to Ohlinger, sought to create obedient, docile, and militaristic subjects— had tainted the public schools in the United States, and if this danger was not stopped "there succumbs the nation." The Americanization movement, therefore, had the added aim of cleansing the schools of all foreign elements, an aim which, at times, became an obsession. In California, a recently purchased, state-authorized music book was found to contain a number of German folk songs. Before the book was distributed to the schools, state authorities, including the Council for National Defense in California, removed the pages with the offending songs. In Indianapolis, the pages in an elementary reader that contained a German poem were glued together so that students would not be exposed to un-American literature. "We leap first," Bierstadt criticized, "and if we look at all, it is after we have stumbled. That is the trouble with Americanization." Bierstadt despised the fanaticism and paranoia of the Americanization movement. "[I]n our dealings with the immigrant," he noted, "we have been unutterably stupid. We have been unsympathetic; we have been lacking in understanding or even in the desire to understand; and we have been cruel." Particularly poignant was Bierstadt's criticism of the Americanizers' vision of the melting pot, "which we alone shall construct...and into which we shall throw any and all ingredients which we regard as foreign." Even Hill, who was sympathetic to the movement but disliked its inefficiency, found some aspects of Americanization a bit extreme. "It is difficult to see," the professor commented, "why true Americanism necessitates on the part of the immigrant the adoption of our foods or our methods of preparing food, as urged by the National Americanization Committee. It is conceivable that one may continue to eat goulash or garlic and forego the pleasures of pie and yet become a true American in mind, heart, and action."[16]

The War against Bilingual Education

The fight against foreign languages in America's schools was an extension of the wartime paranoia and the Americanization movement. The removal of all foreign elements from the school system, particularly non-English languages, was essential because it was through education that the coming generations were molded into upstanding citizens. Moreover, the schools, in this time of crisis, were perceived as central agencies for vast German plots. The "goose-step of Kultur" in America, as Ohlinger called the advance of Prussianized thought, was propagated by professors in American universities, either those of German ancestry or those who had studied in Germany. These scholars, it was argued, were spreading the "German virus" throughout the educational system. To halt this perceived threat, numerous professors in the United States were fired. The University of Michigan, Ohlinger's alma mater, dismissed six instructors in its German Department during the war years. Carl Eggert was the first to be terminated because some of his public statements were interpreted as being pro-German in sentiment. The other five instructors, as historian Clifford Wilcox has noted, "were dismissed not for what they said but for who they were"; four of them were of German heritage, and one, Warren Florer—although not ethnically German—was a staunch proponent of German *Kultur* and connected to the German-American Alliance.[17]

While the war "over here" was fought against foreign influences in the colleges, the war also was waged in the high schools, and German was the primary target. During the war, enrollment in high school German classes waned significantly, and, as in the colleges, numerous teachers of German were dismissed or reassigned. Throughout the nation, secondary German instruction was dropped, with nearly 150 school districts ending their classes. New York City stopped allowing new students to enroll in the high school German program in September, 1918. The director of modern languages in New York, Lawrence A. Wilkins, suggested that Spanish "*cultur*" in the secondary schools would be a welcomed "substitute" for the "dread disease" of German "*Kultur*." The danger, however, was that German teachers, who were already engaged in developing Spanish textbooks, would fill these positions and bring "Germanic Spanish" into the classrooms. Wilkins, therefore, noted that "[w]e want the Spanish language taught in this country by teachers born and trained either in the United States or in a Spanish-speaking land." In addition to the local districts, several states, such as South Dakota, banned the teaching of German in any schools, including in its high schools and colleges. In 1919, Indiana passed legislation that outlawed German studies at the secondary level. The Hoosier law stated that "Latin or any modern foreign language except German" was to be offered in the state's high schools.[18]

The center of the battle against foreign languages, however, was in the elementary schools. The American elementary school was the location of many of the nation's bilingual programs, and its clientele, because of the students' tender age, was thought to be the most susceptible to the dangers of foreign influence. As in the higher levels of the educational system, the German language was the central enemy. In 1919, immigrant Johann Witten reported that "[i]n many places the Germans had a hard time, churches and schools were closed down." In additional to being the language of the "Hun," German was the most prevalent foreign language taught in America's elementary schools just prior to the war; a Bureau of Education report found that, of the cities with bilingual programs that were surveyed, the vast majority offered the language.[19] Although German was often singled out as the most dangerous foe in the educational theater of war, the anti-foreign sentiments that the conflict aroused made their mark on other languages as well. Ironically, this turn of events was the exact opposite of what had begun to occur at the end of the nineteenth century. The arrival of the "new" immigration just prior to the turn of the twentieth century spawned the beginnings of the push for Americanization in the schools. While primarily directed at the newcomers from southern and eastern Europe, the monolingualism of the early Americanization campaigns threatened the bilingual programs of the more-established immigrants, especially the Germans. With the outbreak of the war, however, the tables were turned. The war against the German language in America broadened into a campaign for English-only schools, which challenged the bilingual activities of other foreign-language speakers. The hysteria created by the Great War, in short, was crumbling the polyglot boardinghouse.

Perhaps nowhere was the fight against bilingual education more profound than in the Midwest because it was that region which, during the nineteenth century, had become the cradle of German-English instruction. Even the Midwestern cities where liberal ethnic communities had a great deal of power and influence, such as Milwaukee, Cleveland, Cincinnati, and Indianapolis, could not stem the tide of anti-foreign feelings, particularly with regard to dual-language instruction in the public primary schools. While Commissioner of Education P. P. Claxton saw the value of foreign-language instruction, even during wartime, his Bureau of Education noted that "educators generally look upon the teaching of foreign languages in the lower elementary grades as of very questionable value." In 1916, over 60 percent of the students in Milwaukee's public elementary schools learned German through short daily lessons. In June 1917, just two months after America's declaration of war, the city's school board began the process of removing foreign-language instruction from the primary grades, including the Polish and Italian classes that were relatively new to the school system. Waning enrollments, public pressure, and a superpatriotic atmosphere—one in which Milwaukee's teachers had to sign loyalty oaths—facilitated the complete elimination of foreign-language instruction by 1918. As in Milwaukee, public outrage ensured the demise of Cleveland's German program in June of the same year.[20]

Cincinnati, the virtual birthplace of public German-English schooling, also found its foreign-language instruction under attack. In its seventy-sixth year when the declaration of war came, the city's bilingual program enrolled over fourteen thousand elementary and grammar pupils during the 1916–1917 school year. Yet, the writing was already on the wall, and the director of German-English instruction, Heinrich Fick, certainly recognized it. In his 1917 report to the superintendent, Fick noted that "a steadily increasing opposition to the teaching of the German language made itself felt, becoming acute at once upon the outbreak of the war with Germany. Early in the spring came indications of concerted and determined action looking to the elimination of German study from the elementary schools and the curtailment of it in the high schools." The following year, the enrollment in German dropped to nearly half of its 1916 level. Under intense pressure from the larger society, the city's dual-language course of study, which had been the model for so many bilingual programs throughout the nation, was terminated in February 1918.[21]

In Indianapolis, the initial attack on elementary bilingual education came the month following America's declaration of war. Just prior to the war, however, Indianapolis's German program was flourishing. In 1916, over 8,000 pupils learned the language in the district schools, which was an increase of over sixteen hundred students in seven years. Peter Scherer, the director of modern languages for the city's public schools, noted that the goal of German in the elementary school was "to give the pupil an insight into modern German life, with its customs and manners, its joys and sorrows, and to render him more proficient in every day conversation." Scherer, a liberal-minded club German, subtly expressed his contempt for the war in Europe in his 1916 report to the superintendent. Quoting the psychologist G. Stanley Hall, the director noted that the elementary student of German "will 'feel the spirit of another nation, share its aspirations, understand its points of view, and thus strengthen fraternal bonds and increase the motives of peace and good will.'" Although the program was doing well, the declaration of war with Germany in April led to criticisms of German-English instruction in May when a community organization, the American Rights Committee, accused the German Department of promoting un-American sentiments. This growing public outcry against German instruction facilitated the city's school board to end its bilingual program in January 1918. Although Theodore Stempfel, a leading German American in the city, opposed the resolution, the remainder of the school board declared that "the public schools should teach our boys and girls the principle of one nation, one language, and one flag, and should not assist in perpetuating the language of an alien enemy in our homes and enemy viewpoints in the community."[22]

The war against bilingual education, of course, was not confined to the Midwest. Cities throughout the nation, such as Hoboken, New Jersey, and Buffalo, New York, dropped their elementary German programs during the war years. In heterogeneous cities, popular support for the nation's anti-German crusade undermined dual-language instruction even in those communities that had historically been strongholds for the liberal club Germans. Yet, urban areas, while garnering a great deal of attention, were not the only communities that offered bilingual instruction. All over the country there were small, relatively homogeneous ethnic enclaves that used the public schools to maintain their non-English tongues, and, in such an anti-foreign atmosphere, additional measures were needed to ensure that such "un-American" activities were quashed. To address this threat, states, over twenty of them in all, passed legislation that, to various degrees, hampered bilingual instruction. While most of the language laws went into effect between 1917 and 1921 (see table 5), several states had legislation on the books that mandated English as the language of instruction in the schools even before the outbreak of war. Kansas, for instance, deemed that English was to be the language of the schools in the late 1870s, while Colorado did the same in 1908. Arizona and California's prewar school laws also ensured that English was to be the language of instruction in the states' schools.[23]

Table 5   Language-Restriction Legislation in Elementary Schools by State, 1918–1921[i]

State	Year	Elementary Schools with Foreign-Language Restrictions
Louisiana	1918	Public and Private; Restriction on German Only
North Dakota	1918	Public and Private
South Dakota	1918, 1919, 1921	Public (1918); Public and Private (1919, 1921)
Texas	1918	Public
Alabama	1919	Public and Private
Arkansas	1919	Public and Private
Colorado	1919	Public and Private
Delaware	1919	Public and Private
Iowa	1919	Public and Private
Idaho	1919	Public and Private
Illinois	1919	Public and Private
Indiana	1919	Public and Private; Restriction on German Only
Kansas	1919	Public and Private
Maine	1919	Public and Private; English Required for the Common Subjects Only
Minnesota	1919	Public and Private; English Required for the Common Subjects Only; Foreign Language Limited to One Hour Daily
Nebraska	1919, 1921	Public and Private
Nevada	1919	Public and Private
New Hampshire	1919	Public and Private; English Required for Specified Subjects Only
New Mexico	1919	Public
Ohio	1919	Public and Private; Restriction on German Only
Oklahoma	1919	Public and Private
Oregon	1919	Public and Private
Pennsylvania	1919, 1921	Public and Private; English Required for the Common Subjects Only
West Virginia	1919	Public and Private; English Required for the Common Subjects Only
California	1921	Private
New York	1921	Public and Private; English Required for Specified Subjects Only

Note:

i. The term "private" refers to any non-public school.

Source: I. N. Edwards, "The Legal Status of Foreign Languages in the Schools," *The Elementary School Journal* 24, no. 4 (1923): 272; Henry J. Fletcher, ed., "Recent Legislation Forbidding Teaching of Foreign Languages in Public Schools," *Minnesota Law Review* 4, no. 6 (1920): 449–450.

With regard to the wartime laws, a few states explicitly outlawed German instruction in their schools. In 1918, Louisiana prohibited the use the German language in all of the state's schools, both public and private. The following year, Ohio followed suit, but only forbade German in the elementary and grammar grades. The first of Indiana's 1919 language laws also prohibited German in the elementary and grammar grades, while the second banned German instruction from public secondary schools. Indiana's expulsion of the German language from its public and private elementary schools captured the hysterical mood that engulfed the nation during and immediately after the Great War. Although bilingual instruction had already been eliminated from the capital city's public schools, State Senator Franklin McCray and Lieutenant Governor Edgar D. Bush drafted the anti-German legislation, legislation that was introduced in the state senate on February 17, 1919. The McCray Bill stipulated that English would be the language of instruction in all of Indiana's schools and that German was forbidden in public and private elementary educational institutions; violators of the act would face fines and, potentially, incarceration. The legislation went to the floor of the Indiana House of Representatives on February 25, where the normal procedure of reading a bill "on three separate days" was overruled because it was considered an emergency act. McCray's legislation passed the house in fifteen minutes and was immediately taken to the governor to be signed into law. In less than two hours from its arrival on the house floor, the prohibition on elementary bilingual instruction went into effect.[24]

Unlike those in Louisiana, Ohio, and Indiana, most of the state language laws did not specifically mention German but, nonetheless, made English the official language of instruction in the schools. In 1917, Wisconsin strengthened its language law but retained the elementary schools' ability to give one hour daily lessons in foreign languages. Two years later, Minnesota also passed legislation that limited foreign-language instruction to one hour each day. In 1921, South Dakota mandated that English was to be the language of instruction during the school year, a law that accommodated the Scandinavians' tradition of using the public schools during the off season to maintain their linguistic and religious heritage. Some states, most notably Texas and New Mexico, banned the use of foreign languages in public schools only, while most other states that jumped on the restrictive-legislation bandwagon outlawed foreign languages in both public and private schools. Nebraska's 1919 law was the most notorious, a law that was later found unconstitutional by the U.S. Supreme Court. In

1920, the *Minnesota Law Review* noted that Nebraska's school language act was "the most far reaching legislation enacted upon this subject." The first section of the law stated that "[n]o person, individually or as a teacher, shall, in any private, denominational parochial or public school, teach any subject to any person in any language other than the English language." The second section explicitly prohibited the teaching of a foreign language as a subject before "a pupil...successfully passed the eighth grade." Nebraska took its law quite seriously. In 1920, for instance, a teacher at a parochial school was convicted of teaching reading—in German—to a ten-year-old. This conviction set in motion a legal battle that eventually overturned the excessiveness of Nebraska and other states' language laws.[25]

Although initially directed at German instruction, these state laws naturally had a tremendous impact on other elementary language programs. For instance, the 1919 language laws in the upper Midwest—Iowa, Illinois, and Minnesota—restricted the Scandinavians' ability to maintain their mother tongues in the region. Like many other states, Texas passed restrictive legislation in 1918, legislation that hampered not only German-English instruction, but also the bilingual programs of *tejanos* and Czechs throughout the state. During the late-nineteenth and early-twentieth centuries, the Texas legislature limited bilingual instruction through a series of English-only school laws. This prewar legislation still allowed foreign languages—primarily Spanish, German, and Czech—to be taught as subjects, but the laws were relatively weak and, thus, were widely ignored. During the war years, however, Texan officials who favored Americanizing the Lone Star State took advantage of the anti-foreign atmosphere that the conflict had created. The wartime hysteria not only generated an anxiety about the German element in the state, but also created an uneasiness about Mexican radicalism; after all, the Zimmerman note did suggest that Mexicans could regain their lost territory if they sided with Germany in the international conflict. As the strengthened English-only school bill came before the Texas legislature, it initially excluded only German. A competing bill attempted to outlaw all foreign languages in the state. As a compromise, the 1918 school law banned all non-English languages only from the lower elementary grades, but, unlike earlier legislation, the law included provisions that would punish those who ignored the new statute. As a follow-up to the 1918 legislation, Texas passed a law in 1923 that forced an English-only curriculum on the state's private schools, a move that primarily sought to Americanize the Spanish-speaking population.[26]

Like Spanish instruction in Texas, Japanese-English schooling also became a casualty of the Great War. Prominent in the Territory of Hawaii, as well as in parts of California and Washington, Japanese language schools flourished before 1917. In 1909, for instance, Hawaii had over seventy Japanese schools that enrolled around seven thousand pupils. While many of the schools were operated by religious groups, primarily Buddhist and Christian sects, several were secular in nature. Like the Scandinavians in the upper Midwest, some of Hawaii's Japanese residents used public educational facilities when school was not in session to promote bilingualism among the American-born generations. As on the mainland, World War I ushered in a hysterical drive for Americanization, a form of Americanization that pushed for an English-only education. While the German Americans were the primary targets of the wartime fanaticism throughout the continental United States, in Hawaii the anti-foreign sentiments stirred up by the conflict were directed at the Japanese immigrants and their children. As with the German immigrants, Japanese Americans were accused of disloyalty, especially since Hawaii-born children, while regarded as Americans, were able to retain their Japanese citizenship; in Japan, citizenship emanated from paternal status, not place of birth. When asked about the students attending Japanese language schools, a public school teacher noted shortly after the war that "[i]t is pretty hard to teach American ideals to a child who does his thinking in Japanese." Another teacher responded, "The Japanese schools, under cover of religious instruction, teach the children loyalty to their Emperor and country. The Japanese language schools must go, if we are to teach the young Japanese to become Americans." This fear of divided loyalties merged with the anti-Asian, racialist form of nativism that became particularly acute during the Progressive period. The Japanese, like the Chinese before them, were seen as inferior to whites and, thus, were targeted for discrimination. By the early 1920s, Japanese language schools were shackled not only in Hawaii, but in California as well; both required that the schools promote "Americanism" and mandated that the teachers be certified by territory or state officials. Although prominent leaders in Washington also sought to shut down the state's Japanese schools, these legislative manifestations of nativism failed to garner support, and many language schools in the state persevered.[27]

While the state (and territory) language laws that were enacted all over the United States during and after the war were largely a reflection of the anti-foreign, particularly anti-German, atmosphere that the conflict had generated, the laws also sought to Americanize the small ethnic communities that, potentially, would be shielded from the sort of pro-war public opinion that was more common in the heterogeneous cities. Although perhaps not a direct result of the state language laws, the Great War, nevertheless, did tend to diminish many foreigners' ethnic identities and began to draw them out of their enclaves. While Milwaukee was a large, multi-ethnic city, one that succumbed to the public pressure to end its public German-language instruction, its north side was very much a Germanic island prior to the war, an island where, as the prominent language scholar Robert F. Roeming stated, "all forms of social and community life could...be conducted in German." Born in 1912, Roeming was an elementary student at Milwaukee's St. Marcus Schule when the war began. Mornings at the school were devoted to German and the afternoon lessons were taught in English. Although it promoted *Kultur*, the parochial school, according to Roeming, was certainly not a tool of the German empire; in fact, the liberal club Germans and Lutherans of the city had generated a "culture...[that] was opposed to the political philosophy of Imperial Germany." Decades after the conflict, Roeming, then a professor at the University of Wisconsin-Milwaukee, recalled that World War I had the effect of bursting the isolationist, ethnic bubble that surrounded Milwaukee's Germanic north side. The urban expansion and economic opportunities that emerged during and after the war made the homogeneous north side more connected to other areas of the city, as well as to the larger nation. Moreover, the drafted soldiers, venturing out of their communities for the first time, caught a glimpse of the wider world, and upon their return "could not very easily reorient themselves to find their way back into the socio-psychological German ethnic enclave."[28] The war and the wartime measures, therefore, were both directly and indirectly breaking down ethnic homogeneity and ushering in a new, more-cosmopolitan form of American unity.

In addition to disrupting ethnic isolation, the state language laws also filled a psychological need among many Americans. Having been left somewhat impotent on the home front, the war "over here" seemed to be, partially, an outlet for the anger many felt but could not unleash on the battlefield. As the anti-German school bill passed in Indiana, for example, a state legislator expressed his rage that the law did not outlaw all foreign languages. "We want no little Germanys," the state official noted, "nor do we want any little Italys or other groups." The Woman's Auxiliary of the Rainbow Regiment Cheer

Association in Indianapolis, an organization made up of female relatives of soldiers, also resented German-language instruction, a resentment that was personal and intimate. While their husbands, sons, and brothers remained in Germany as part of the occupation, the members of the association wanted all vestiges of the culture that their loved ones encountered "over there" to be eliminated from American society before they came home. German-English schooling, the organization resolved, was "a failure to keep faith with our heroic dead and with our men who, returning, will expect to find the state prepared to carry on ideals for which they fought." Hinting at the need for revenge, the Rainbow Regiment also noted that "the German language in the elementary schools of the state will be a tool in the hands of the German propagandists who are seeking to bring about a soft peace with Germany." Of course, there would be no soft peace with Germany; the "peace," in fact, was so harsh that it would help usher in another world war.[29] The conclusion of the war against bilingual education also had a harsh ending, an ending that witnessed the near total destruction of America's polyglot boardinghouse; by the 1920s, only scattered ruins were left.

8

Rebuilding the Boardinghouse: The Interwar Years

After the Great War, the polyglot boardinghouse, like the French countryside "over there," had been torn to sheds and was left in shambles. Nevertheless, portions of the boardinghouse remained—just scattered ruins really, but enough foreign-language rubble to use as materials for reconstructing at least sections of multilingual America, a multilingual America that was partially supported by bilingual education. The notion that dual-language instruction did not simply stop after World War I runs counter to the assertions of language scholars who, particularly in the 1960s and 1970s, began to explore the history of bilingual education in the United States. In 1971, Theodore Andersson, a professor of language education at the University of Texas, claimed that "[b]ilingual schooling...disappeared from the United States scene between 1920 and 1963." Although Andersson was a leading authority on the history of dual-language instruction in America, his perspective underestimated the bilingual activities in the nation's schools during the 1920s and 1930s, decades that were pivotal in the development of the modern bilingual education movement of the 1960s, a movement in which Andersson was a central figure.[1]

The interwar period was an era of extremes, from exceptional prosperity to the dark depths of an economic depression. During the 1920s, the Dow Jones Industrial Average increased fivefold, registering over 380 points just before the crash. During the 1928 presidential campaign, Herbert Hoover expressed his optimism that flowed from the economic boom. "We," the future president told a New York audience, "are nearer today to the ideal of the abolition of poverty...than ever before in any land." This coming utopia, however, was a façade that masked deep inequalities. The financial gains of the 1920s did not reach all layers of the society. Ludwig Dilger, a working-class German immigrant, wrote in the mid-1920s that the media "boast a lot here about the great prosperity of the people, but wages are low, except for construction workers, and the cost of living is terribly high." He concluded that "[e]very day the rich get richer and the poor get poorer."[2]

Although the prosperity of the 1920s did not touch everyone, the crash and the deepening depression of the 1930s exacerbated the plight of those already struggling and unraveled the financial security of many of those who had participated in the economic boom. By 1933, the year in which Franklin Delano Roosevelt took office, conservative estimates suggested that a quarter of Americans were unemployed. In his inaugural address, President Roosevelt tried to boost Americans' confidence but bluntly noted that "[v]alues have shrunken to fantastic levels; taxes have risen; our ability to pay has fallen...; the withered leaves of industrial enterprise lie on every side; farmers find no markets for their produce; [and] the savings of many years in thousands of families are gone." Dilger, writing from St. Louis the year before FDR's inauguration, put forth a similarly bleak evaluation of the financial crisis in the United States. "We have between 10 and 12 million unemployed here. And the taxes keep going up. Thousands have lost their property because they weren't able to raise the money for the high taxes." Faced with this uncertainty, as one of historian Studs Terkel's interviewees—Pauline Kael—noted, "Families had totally broken down." Fathers, Kael told Terkel, "had wandered off in disgrace because they couldn't support their families. Other fathers had killed themselves, so the family could have the insurance."[3]

The Great War, the economic prosperity of the 1920s, and the Great Depression certainly were pivotal events in American history, but they did not usher in a complete break with the past. For generations, historians—most prominently Richard Hofstadter—suggested that progressivism ended with World War I. But, more recent scholarship has emphasized the continuation of progressive ideals in the interwar years. Morton Keller, for instance, has noted that "as our historical perspective lengthens, we can see more clearly that the notion that World War I sharply split the early twentieth century into two periods is a strained and artificial one." The "age of reform," as Hofstadter has called it, persisted through the "Roaring Twenties" and, ultimately, helped shape the New Deal. The postwar years witnessed numerous successful reform initiatives, including prohibition and women's suffrage, both of which became constitutional amendments. In addition, it was during and after the war that intellectuals such as Horace Kallen and Randolph Bourne railed against the homogenizing aspects of Americanization and called for a pluralistic or cosmopolitan view of American culture, thus expanding on the work of humanitarian progressives such as Jane Addams. FDR's New Deal, of course, continued the reformist impulse of progressivism and combined a humanitarian care for the downtrodden with a desire for more governmental regulation and efficiency. Faced with an unprecedented financial meltdown, Roosevelt signed the National Recovery Act during his first year in office and the Social Security Act two years later. The New Deal programs and agencies—including the Works Progress Administration, the Civilian Conservation Corps, and the National Youth Administration—put many struggling individuals back to work and gave some of them vocational training, including secondary students. These elements—World War I, the economic boom and bust, and the continuation of progressivism—as well as others (such cultural and demographic shifts) directly impacted schooling and bilingual education during the interwar years.[4]

Interwar Schooling

The 1920s witnessed a massive expansion in the realm of schooling. Enrollments, particularly among high school pupils, soared, and city after city increased spending on education to keep pace with the growing student population. During the depression, enrollments rose even more dramatically because, for adolescents, finding employment became much more difficult; child labor laws and the economic downturn, which frequently hit young workers the hardest, left teenagers with few options, except for high school. At the secondary level, the number of students jumped from almost 4 million in 1929 to over 5.5 million five years later; thus, the high school began to evolve into a sort of holding institution for some youngsters, an institution used to keep them out of the job market. This exponential growth in the midst of the Great Depression posed a serious problem for school districts across the nation. Retrenchment became the new reality; teachers' salaries were slashed, and educational components that were seen as non-essential—often labeled "fads and frills"—were slated for elimination. Schools throughout the nation reduced or eliminated programs such as art, music, vocational training, and evening classes during the early 1930s. In Detroit, a city particularly hard hit by the depression, salaries for teachers were cut by nearly 22 percent; unlike many other areas, Detroit, through a powerful coalition of labor and educational advocates, managed to maintain all of its public school programs, even though some of those programs were attacked as frivolous during the roughest years of the depression.[5]

The assault on educational "frills" was not only a tough budgetary decision for financially hard times; it also was a partial rejection of progressivism because much of what was deemed unnecessary during the retrenchment of the 1930s was the making of educational reformers. Despite the attacks on educational "frills," progressive educational endeavors continued throughout the interwar years. Progressivism in the realm of education, however, was multifaceted, and, instead of a singular movement, a variety of educational "interest groups" sought to influence the direction of the curriculum. The child-centered wing of progressive education continued to make gains during the 1920s and 1930s. The Progressive Education Association, the national organization of this arm of progressivism, was founded (1919) and flourished throughout the interwar period, and child-centered activities continued to filter into America's classrooms, particularly at the elementary level. Although secondary instructors tended to resist child-centered techniques, elementary teachers, attracted to progressive methods, usually adopted such practices piecemeal, creating a sort of hybrid of traditional and student-centered pedagogy.[6]

While many of the educational "frills" that came under attack were clearly intended to make the curriculum more hands-on and in tune with students' interests—and, thus, fell into the orbit of the child-centered educators—some of the programs slated for elimination, such as vocational training, were promoted by both child-centered and social-efficiency progressives, although for vastly different reasons. Social-efficiency advocates saw vocational education not as a means of making schooling more engaging, but as a way to "efficiently" direct the new student population, which tended to come from the working classes, into its probable career path. Although threatened by retrenchment policies, vocational training endured during the economic turmoil and continued to receive substantial federal allocations. In addition, New Deal agencies, such as the National Youth Administration, stepped up vocational training for working-class youngsters outside of the school setting, somewhat to the chagrin of educational leaders who saw such a move as infringing upon their turf. The social-efficiency wing of progressive education thrived during the interwar period because it captured the tenor of the times. Throughout the 1920s, the corporate model on which socially efficient progressivism was based was, naturally, in tune with the business leaders who were prospering. The desire to remake school systems into efficiently running corporations continued unabated during the "Roaring Twenties," a point confirmed by the continued popularity of "school surveys." In 1922, the survey of the El Paso schools, conducted by veteran surveyor Paul Horn, made numerous recommendations for making the schools more socially efficient, including ensuring that Chicano students "receive instruction....suited to...[their] needs and capacities," such as "handwork," "cooking," "sewing," and "manual training." Moreover, the hard-line business orientation of social-efficiency progressives also fit with the retrenchment policies of the 1930s. As corporate leaders were called into action during the depression to "save" the schools, their fat-trimming measures were, by and large, in line with the recommendations of progressives who wanted to make the schools more "efficient."[7]

There were, of course, other educational progressives who rejected much of what the social-efficiency and child-centered leaders championed. Often referred to as social-reconstructionist progressives, early humanitarian reformers such as Jane Addams were leery of socially efficient education, noting in 1902 that "our schools have become insensibly commercialized." What disturbed many critics of educational efficiency was its undemocratic nature. Both Addams and John Dewey envisioned schools as democratic communities, and, as Dewey noted, "[a] democracy is more than a form of government; it is primarily a mode of associated living, of conjoint communicated experience." During the 1930s, when the social-reconstructionist critiques of business efficiency resonated because of the depression, George Counts of Teachers College expanded on Dewey's notion of democracy by noting that a "society fashioned in harmony with the American democratic tradition would combat all forces tending to produce social distinctions and classes; repress every form of privilege and economic parasitism; manifest a tender regard for the weak, the ignorant, and the unfortunate;...[and] strive for genuine equality of opportunity among all races, sects, and occupations." Counts and other socially conscious educators, such as Dewey, often outlined their reform agenda in the journal *The Social Frontier*, which began publication in 1934. These social reconstructionists hoped to use education not as a vehicle for maintaining the status quo—which was essentially what the social-efficiency progressives did by fitting the children into the existing social and economic hierarchies—but as a means building "a new social order," one that was more just and humane.[8]

Counts also rejected the tenets of the child-centered progressives. During the early 1930s, he addressed the Progressive Education Association and stated that "[i]f an educational movement...calls itself progressive, it must have orientation; it must possess direction." "The weakness of Progressive Education," Counts continued, "lies in the fact that it has elaborated no theory of social welfare, unless it be that of anarchy or extreme individualism." The professor did not limit his critique of schooling to the individualist underpinnings of the child-centeredists, however. He also brought forth a challenge to the values of modern America, the new public morality that had made its way into the schools throughout the Progressive era. In order to navigate an increasingly complex society, ethical flexibility was at a premium, and the "business men," as Addams stated, who Americans "so tremendously admire" were the model of this moral plasticity. To prepare students for "modern living," as one superintendent in Indianapolis called it, schools needed to help pupils develop the sort of dispositions that were required for interpersonal communication, which often translated into a focus on the improvement of "personality." A businessman himself, Sinclair Lewis's fictional George F. Babbitt saw the virtue of correspondence courses that sought to teach students "[h]ow to create a strong personality" in order to pilot modern economic realities. Counts was disturbed by this direction the schools were taking, noting that it was a "fallacy" to think "in a dynamic society like ours the major responsibility of education is to prepare the individual to adjust himself to social change." The businessman's notion that "the individual who is to live and thrive in this world must possess an agile mind, be bound by no deep loyalties, hold all conclusions and values tentatively, and be ready on a moment's notice to make even fundamental shifts in outlook and philosophy" was, according to Counts, utterly ridiculous. "Under such a conception of life and society," the professor insightfully noted, "education can only bow down before the gods of chance and reflect the drift of the social order."[9] Social reconstructionists, by contrast, called for Americans to take a humane and principled stand on social issues. The rise of social reconstructionism, along with the other strands of progressive education and the conditions of interwar schooling, reworked the nature of bilingual instruction in the United States.

Minor Reconstruction among the Europeans Immigrants

While the Great War "over here" sought to end America's polyglottism, bilingualism in the United States, even in its schools, could not be extinguished in such a short period of time; public schools and educators can be (and often are) resistant to abrupt change. The German immigrants, at whom many of the wartime English-only policies were directed, continued to speak their native language. During the 1919–1920 school year, for instance, Dolly Holliday, a nineteen-year-old educator, took a teaching position in a German-Russian community on the northern plains and quickly discovered that there was a "language barrier" between her and the students. Church services and many local conversations were conducted in the residents' mother tongue, and, therefore, Holliday, for the sake of a friendship, "learned a little German," while her friend "learned a little English." (Ultimately, the cultural clash that Holliday encountered in the German-Russian town prompted her seek an appointment in a different community for the following year.) The rural towns on the northern plains were not the only localities in which *Deutsch* survived. World War I did not automatically make Anglophones out of America's Germans. Analyzing the 1940 census, historian Walter D. Kamphoefner has found that among the aging second-generation German population "there was no evidence of a sudden jump in English speaking as a result of World War I." In fact, the use of German among this cohort largely continued unabated in the United States, particularly in rural areas.[10]

Although World War I did not stamp out bilingualism, it did dramatically reduce bilingual education among the Germans throughout the nation. Nevertheless, dual-language instruction endured. Julius K. Hoffmann, the pastor during the conflict for Baltimore's Zion Church, "maintained classes on two afternoons each week, for those who sought them, even in those hectic days when it was considered a criminal offense to speak German." Although the church's German-language school began in the eighteenth century, it gained renown in the following century under the leadership of the charismatic Heinrich Scheib. The German-English program of the city's public schools lessened Germans' reliance upon the parochial schools, but the Great War stunted such public forms of bilingual instruction and, once again, the responsibility for language maintenance fell largely upon the private sector. The Zion Church School, therefore, reopened in the late 1920s with the goal of, as teacher Else Hoffmann noted, "instill[ing] into the child a love for German speech and song." Throughout the 1930s, the German-language school had an enrollment of about two hundred students each year, and the pupils, many of whom were American born, gained "a fair knowledge of reading, writing...[and] grammar." The students in the early grades partially learned the language through the textbook *Deutsch für Anfänger*, a "cheerfully red" "little volume" written by the school's principal, Elsa S. Conradi.[11]

During the interwar years, bilingual education also was a public endeavor in some areas of the country. Although many states outlawed bilingual education, some, such as Wisconsin and Minnesota, continued to allow for a modicum of foreign-language instruction in the elementary schools. These two states in the upper Midwest permitted daily one-hour lessons in languages other than English. But, bilingual education also persisted elsewhere, largely because it was a necessity for student understanding. Because of the restrictive state laws, public dual-language instruction took on a covert form not unlike the bilingual activities of the Chinese and Native Americans in the English-only classrooms of the nineteenth century. Esther Vaagen, a teacher in rural North Dakota, began her career in 1915 and learned that instruction in German was an essential pedagogical strategy. After the war, Vaagen noted, "German was forbidden around there," but the immigrant students, representing many nationalities, "could understand German even if they weren't true Germans." "Although it was forbidden," the rural teacher stated, "I told them in English, then in German, and then in English and it was surprising how fast those little tots caught on." While it was somewhat unusual that Vaagen explicitly confessed to skirting the law, this covert use of bilingual instruction was probably not an isolated practice. In a 1917 survey of teachers in Illinois, L. D. Coffman found that a large portion of the state's teaching corps—about 20 percent—came from foreign-language-speaking homes. These bilingual teachers, most of whom had Germanic origins, likely aided—albeit somewhat clandestinely— their immigrant students in languages other than English when they failed to understand, as Vaagen did in rural North Dakota.[12]

This covert form of bilingual education among the older immigrant groups largely became unnecessary after two important decisions by the U.S. Supreme Court in the 1920s, decisions that also facilitated the reestablishment of private language schools, not unlike the Zion Church School in Baltimore. In 1923, Justice James C. McReynolds delivered the Court's majority opinion in the *Meyer v. Nebraska* case. Nebraska's 1919 school law made it a criminal offense to "teach any subject to any person in any language other than the English language" before the ninth grade. In Nebraska's Hamilton County District Court, the plaintiff, a parochial school teacher, was convicted of violating the legislation because he "taught the subject of reading in the German language to Raymond Parpart, a child of ten years." Although the decision was upheld by the state's highest legal body, the U.S. Supreme Court determined that the Nebraska law violated the "liberty guaranteed" by the Fourteenth Amendment. Parents, the Court noted, had the right to send their children to any school of their choosing, and the Nebraska school law had undermined this right. Justice McReynolds concluded by stating, "No emergency has arisen which renders knowledge by a child of some language other than English so clearly harmful as to justify its inhibition with the consequent infringement of rights long freely enjoyed."[13]

Like the Nebraska law, Oregon's 1922 compulsory school act sought to Americanize the state's foreign element. Supported by the state's powerful Ku Klux Klan, the law required that every child attend *public* school, thus undercutting the influence of ethnic private and parochial education. In the 1925 *Pierce v. Society of Sisters* decision, McReynolds, again handing down the Supreme Court's opinion, drew upon the *Meyer* precedent and noted that it was "entirely plain that the Act of 1922 unreasonably interferes with the liberty of parents and guardians to direct the upbringing and education of children under their control."[14] The *Meyer* and *Pierce* cases, therefore, undermined the restrictive state laws that hampered both public and private bilingual instruction. Although the Supreme Court had cleared the way for the resumption of bilingual education, a number of factors came together after World War I that shifted dual-language instruction away from European groups and made it a pertinent issue for other foreign-language speakers in the United States.

Immigration restriction was one of the central factors that lessened the importance of bilingual education for European groups. By the turn of the twentieth century, the tide of the "older" immigrant groups, those who had been most involved in public foreign-language instruction, had receded, only to be replaced by a "new" immigration from southern and eastern Europe. War or the threat of war traditionally has been a catalyst for national unity in America, a unity that frequently takes on racial overtones, and World War I was no exception. Due to a fear of the new racial immigrant stock, particularly its "inferiority" and supposed radicalism, the U.S. Congress enacted a law on February 5, 1917, requiring that immigrants be literate, a stipulation that excluded a multitude of would-be Americans. Nativists called for more far-reaching legislation to halt the foreign threat in the United States, legislation that reflected the racialist attitude of the time. In 1921, Prescott F. Hall, secretary of the Immigration Restriction League, argued that the inferior European "races" had not supported the war effort, including the newer German immigrants, whom Hall believed were of "Alpine" rather than "Nordic" stock. "The World War," Hall concluded, demonstrated "that the superficial changes constituting 'Americanization' were entirely inadequate to affect the hereditary tendencies of generations." The position of Hall and other restriction advocates was parodied in Lewis's 1922 novel *Babbitt*. "Thank the Lord," the nativist character (with a Slavic surname) asserted, "we're putting a limit on immigration. These Dagoes and Hunkies have got to learn that this is a white man's country, and they ain't wanted here. When we've assimilated the foreigners we got here now and learned 'em the principles of Americanism and turned 'em into regular folks, why then maybe we'll let in a few more."[15]

The racial nativists' designs were codified in 1924 with the passage of the Johnson-Reed Act. By 1921, the federal government had established a quota system to regulate immigration, but the quotas were based on the 1910 census, which, although allowing only "3 percent of the number of persons of such nationality who were resident in…1910" and, thus, greatly reducing the flow of foreigners into the United States, did not directly target the "inferior" immigrants that racialists feared. By contrast, "[u]nder the act of 1924," the U.S. Congress noted, "the number of each nationality who may be admitted annually is limited to 2 percent of the population of such nationality resident in the United States according to the census of 1890." Because the "new" immigration just began to materialize in 1890, the Immigration Act of 1924, which was signed into law by President Calvin Coolidge, drastically reduced the number of southern and eastern Europeans migrating to America. The annual quotas for Italy and Russia in 1924, for instance, were set at 3,845 and 2,248 respectively, while the limit on German immigration was 51,227.[16] Compared to the laissez-faire policies of the nineteenth century, the Immigration Restriction Act virtually shut down the flow of European foreign-language speakers into America, immigrants who might have required or—if they attained enough political clout through numerical dominance—demanded bilingual instruction in the nation's public schools.

Bilingual and foreign-language instruction among several European groups, while certainly lessening and not nearly as ubiquitous as German education prior to the war, did make a few inroads in America's schools during the interwar period. Immigration restriction certainly did halt the influx of newcomers, but, as the Italian immigrant Leonard Covello noted, the legislation had not "disposed of the need for a further discussion of immigrant problems and of programs for the foreign-born." After all, Covello pointed out that nearly a third of the U.S. population in 1930 was either foreign-born or of the second generation; in New York City, where Covello was a high school principal, over 70 percent of the inhabitants were immigrants and their children. With the enormous amount of foreign-language speakers in America, bilingual education, even in a period as marked by rabid xenophobia as the 1920s, was inevitable, at least on a limited scale. Language scholar Joshua Fishman, for example, has noted that "the twenties may properly be referred to as the 'golden age' of Hungarian-American language maintenance efforts." The postwar treaties, which cut away nearly two-thirds of the Hungarians' ancestral land, "tended to solidify and strengthen...[the Hungarian community's] ethnic life in the United States." The renewed commitment to their ethnic identity coupled with the economic prosperity of the decade facilitated and financed the intensification of bilingual instruction among Hungarian Americans. Primarily private-sector endeavors, bilingual and foreign-language studies became common in the parochial and independent schools administered by the Magyars, leading Fishman to conclude that "virtually every Hungarian-American child went to a Hungarian language school, except in the rural and mining communities where the more closely knit family life served the same language maintenance function."[17]

The Great Depression largely halted the Hungarian-English educational endeavors—even those supported by the Catholic Church—as more and more Magyars in America lost their jobs, but the 1930s, despite the economic crisis, did witness some growth in the realm of bilingual instruction. As the retrenchment policies began to ease after the middle of the decade, elementary instruction in French began to flourish throughout Louisiana, a state that, as of 1940, had over four hundred thousand residents who spoke the language. In 1937, Louisiana's board of education resolved to allow for foreign-language instruction in the state's public elementary schools. By 1939, the policy had been widely implemented, with around two hundred educators teaching over six thousand elementary students the French language. French, however, was not the only language making gains during the interwar period. Instruction in Italian and Hebrew, particularly at the secondary level, increased throughout the 1920s and 1930s in cities such as New York. With the growing popularity of Italian instruction among Italian Americans, several junior high schools added the language, as did some elementary schools, most notably those in San Francisco. By the late 1930s, Covello, a longtime official for the Italian Teachers Association and a former head of the Italian Department in a New York high school, began to make the case for full-fledged bilingual instruction. Drawing on the tenets of both the child-centered and humanitarian wings of progressive education, Covello called for a "transitional use" of dual-language instruction for immigrant students because for "the non-English-speaking groups, it is futile to use only the English language"; bilingual education, therefore, was a child-centered pedagogy because it connected the curriculum to the student's existing knowledge base. But, dual-language instruction also was a social good because such a technique helped quell the generational conflicts that were so common between immigrant parents and their rapidly Americanizing children by officially validated the parents' heritage in the school setting.[18]

Although the interwar years witnessed some minor rebuilding of the boardinghouse among European groups, the period also saw the development of demographic, cultural, and educational changes that worked against a dramatic reconstruction. New migration patterns within the United Stated, for instance, facilitated a reworking of the nature of American ethnic identification. World War I generated a demand for labor, and the conflict and immigration restriction shut down the surge of European newcomers—traditionally the source of cheap employees for the nation's industries—which compounded the shortage. In the early decades of the twentieth century, therefore, southern Blacks began their "Great Migration" to northern cities, where "steel mills," as the Nobel-Prize-winning novelist Toni Morrison has noted, "were begging for workers." The Great War, which was fought for "democracy," made salient the contradictions between American ideals and the reality of discrimination in the South. Emboldened by the war and motivated by racism, poor-quality schooling, unemployment, and the semi-servitude of the sharecropping system, African Americans went north en masse. "In shifts, lots, batches, mixed in with other families," Morrison has written, the southern Blacks "migrated...[to] where there were mines and millwork." By the 1930s, about two million African Americans had made the trip northward.[19]

The migration of southern Blacks disrupted the traditional ethnic composition of northern cities and facilitated a redefining of ethnic identity in America. With the presence of large numbers of African Americans in their midst, immigrants, who had once defined themselves through their languages, nationalities, religions, and Old-World regional affiliations, increasingly came to see themselves as "white." This partial replacing of immigrants' ethnic identities with a simplistic racial vision of themselves was a process that occurred even before the Great Migration. In the 1880s, for instance, Matthias Dorgathen, a German immigrant who had settled in the coal mining districts of Ohio, encountered the American "*Pläklinge* [black-legs]" who were "real riffraff" because "when the others are striking for their fair rights then they do the work." "*Pläck* means *schwarz* [black] in English," Dorgathen told his family back in Germany, "so they compare the whites who work here when there's a strike with the Negroes." Dorgathen insinuated that even white scab laborers were considered "Negroes," but the categories of black and white workers, as opposed to ethnic classifications, became even more pronounced as the southern African Americans made their journey to the North. In Philadelphia's ethnic neighborhoods, historian Russell A. Kazal has argued that the war, nativism, nationalism, and the arrival of large numbers of Blacks encouraged many immigrants to "rework their identities." With African Americans living in close proximity to Irish, German, and other ethnic neighborhoods, the inhabitants "increasingly saw themselves as sharing a common white identity with such neighbors." In addition, some immigrants, even the southern and eastern Europeans, began to partially shed their ethnic identities in the 1930s as they bought into Roosevelt's brand of "civic nationalism," a patriotism that, unlike earlier expressions, seemed more inclusive and, therefore, held out the privileges of whiteness to the "new" wave of immigrants.[20]

The intercultural education movement also contributed to this general decline of ethnic identification among the European immigrants. Gaining prominence during the interwar period, intercultural education was, officially, billed as a progressive means of embracing America's pluralism in the classroom. One of the central leaders of the movement, Rachel Davis DuBois, hoped that intercultural education would lead to a greater sense of respect for the religious, ethnic, and racial diversity in the United States. But, as the movement gained force during the 1930s and 1940s and as more progressive educators jumped on the bandwagon, the aims of intercultural education went from teaching about and, thus, respecting differences to stressing American unity and cohesion as a means of generating tolerance. Covello, although more pluralistic in his leanings than some, recognized and supported the "assimilation" aim of the movement, noting that, ultimately, the goal of education was "in creating an educated, happy, homogenous people" "[I]ntercultural education," as historian Nicholas V. Montalto has convincingly argued, "was not so much the antithesis of Americanization…as it was a more 'scientific' Americanization, a modified expression of the same impulse. The goal remained the same, only the methodology had changed."[21] As the European immigrant groups began to partially assimilate—as a result of the war, immigration restriction, a new racial ("white") awareness, stepped-up nationalism, and remodeled Americanization programs—bilingual instruction, although making some gains in the interwar period, became less of an ambition for many of these foreign-language speakers.

Rebuilding the Boardinghouse: The Non-Europeans

Other "foreigners," particularly those classified as non-white, did not cast off their ethnic affiliations so readily during the interwar period. The Spanish-speaking populations in the United States, for instance, often clung tightly to their cultural and linguistic identities. In New Mexico, many inhabitants saw themselves as "Spanish Americans," as opposed to having Mexican origins, and took great pride in this identity as well as the language that accompanied it. Like the old Dons after the American conquest of the Southwest, New Mexicans bolstered the notion that they were of Spanish heritage, a heritage that some historians view as a "false consciousness" and that others, such as historian John M. Nieto-Phillips, suggest—in a more nuanced manner—was "invented." The scholar George I. Sanchez did much to sustain this Spanish-American identification. In his study of *nuevomexicanos*, titled *Forgotten People*, Sanchez referred to New Mexico's inhabitants as "[t]he descendants of the Spanish colonists" and insisted that these "Americans of Spanish descent" had been "left stranded upon these inland shores" after the "glories of the Conquest." Nevertheless, Sanchez merely reflected what the population already believed. Frances Esquibel, for instance, came from a long line of proud New Mexicans on her mother's side, and the maternal family and others "in the Santa Fe area...were oriented toward a Spanish rather than a Mexican heritage." This distinction between *nuevomexicanos* and *mexicanos* was also linguistic in nature. "Individuals," Esquibel wrote, "were identified as Mexican (as opposed to New Mexican) by the way they spoke Spanish—with what I identified as a singsong quality."[22]

With its strong attachment to Spanish culture and language, New Mexico continued to participate in public bilingual schooling even after the hysterical Americanization movement that World War I facilitated. This continuation of dual-language instruction not only demonstrates that Andersson's notion that bilingual education "disappeared" after the war is exaggerated, but also suggests that such a notion was somewhat Eurocentric since that perspective seems not to have taken into account the Spanish speakers of New Mexico. Adhering to the tenets of the *hispanidad* movement, New Mexicans promoted bilingual schooling throughout the Progressive period, particularly under Superintendent of Public Instruction Amando Chaves and his Spanish-speaking successors. In 1905 and again in 1907, however, Anglos—Hiram Hadley and James E. Clark, respectively—were appointed to this influential post; in 1912, the superintendent of public instruction became an elected position. Hadley and Clark were outspoken opponents of bilingual schooling and hoped, during their tenures, to Americanize the New Mexican public schools. Such designs, however, did not sit well with large segments of the state's Spanish-speaking population. In response to the drive for English-only schools, the territorial legislature, with its large numbers of Spanish-speakers, established the Spanish-American Normal School in 1909. With the explicit aim of educating bilingual teachers, the Spanish-American Normal School, which also offered vocational training, endured in New Mexico for over three decades.[23]

Although after the first decade of the twentieth century the Anglo educational leaders at the state level began to push for more English-only policies, many prominent *nuevomexicanas* resisted such courses of action. Aurora Lucero and her cousin, Adelina Otero-Warren, tirelessly promoted dual-language instruction throughout the new state. Otero-Warren, a county superintendent of education until 1929, developed a popular non-transitional bilingual curriculum that sought to make students fluent in both Spanish and English; her bilingual methods were so uniformly esteemed that the Taos County Teachers Association resolved in the late 1930s that they should be mandated by the state. As in other parts of the nation, the state educational

leaders had to contend with local practice, and this tradition of localism was what, at least partially, kept the English-only advocates from
fully attaining their goals. New Mexican education in the 1930s was
an intensely local endeavor, particularly in the rural areas. In his 1934
dissertation, Sanchez, that great advocate of improved education for
the Spanish-speaking children of New Mexico, noted the presence
of an "excessive amount of decentralization...in the rural schools."
In the rural areas, the county school superintendent was elected by
locals, and the board of education was appointed by a district judge,
who also was elected by community members. This system, in turn,
ensured that local values made their way into the public schools,
including the Spanish language. About 50 percent of the state's population was Spanish-speaking, and over 50 percent of the teaching
corps in some counties—such as Mora, Rio Arriba, San Miguel,
Taos, and Valencia—were "Spanish Americans," thus perpetuating
the use of bilingual instruction in the New Mexican public schools.
In areas that had predominantly Anglo teachers, bilingual education,
of course, was less available, although this was not always the case. In
the 1930s, the University of New Mexico's San Jose Training School,
situated in a county in which only 16 percent of the teaching corps
was Spanish-speaking, utilized a bilingual method in the elementary
grades, a method that, according to school officials, yielded "encouraging" results. In most areas in which Anglo teachers predominated
bilingual education was discouraged and, in many instances, strictly
prohibited. When Esquibel began kindergarten in Eddy County during the mid-1930s, she remembered that "Spanish was absolutely forbidden" in her segregated school. A few years before young Esquibel
began her school career, Sanchez found that Eddy County employed
no Spanish-speaking teachers. The teachers at Esquibel's school were
English-speaking Anglos, and the principal, who also served as a
teacher, was, as described by Esquibel, "a Nordic type," who was neither "warm [n]or solicitous" toward her young charge.[24]

While Esquibel's maternal family was New Mexican, her father
was a Mexican immigrant who, like the *nuevomexicanos*, took
great pride in his cultural and linguistic identity, and who, like the
African Americans, was part of a great migration northward in the
early decades of the twentieth century. Between 1900 and 1930, *los
mexicanos* came to the United States in vast numbers, numbers that
help maintain a vibrant Chicano culture throughout the Southwest.
While the European emigrants increasingly encountered restrictions
on their admittance into America, newcomers from Latin America

(and Canada) were exempt from the quotas set by the Immigration Act of 1924. Besides, many Mexican immigrants entered the United States without going through any formal or bureaucratic channels; they simply crossed the Rio Grande. The Mexican newcomers came to America for a variety of reasons, but one of the most salient was the economic opportunities in *"el Norte."* By the late-nineteenth century, a scarcity of agricultural land for small farmers and an oppressive sharecropping system in Mexico had generated a general movement to urban areas in the nation. But, employment in the Mexican cities was often unstable and yielded poor earnings. By comparison, *los Estados Unidos* offered steady work that paid higher wages, and, therefore, Mexicans crossed the border, officially and unofficially, with the hope of acquiring economic security. "Mexican labor," Julian Nava, whose parents were immigrants themselves, wrote, "was valuable to the United States. After the First World War ended, no one bothered to count Mexicans crossing the border." Like many Mexican migrants, Esquibel's father, Teodoro, came to Texas in pursuit of financial betterment and found employment with an American construction company, an industry that, along with agriculture, hired numerous *mexicanos.*[25]

While many Mexicans were lured to America by its economic opportunities, these immigrants were simultaneously being forced out of their homeland by the violence generated during the Mexican Revolution. The economic turmoil that fueled the migration northward contributed to a conflict with Porfirio Díaz's government in 1910. After Díaz was overthrown in 1911, the nation was thrust into a violent contest for power, a contest that witnessed numerous factions struggling for control of the country. During the bloody civil war, Mexicans migrated to the United States en masse as a means of escaping the ever-present danger. Nava, the future ambassador to Mexico under Jimmy Carter, remembered stories about his parents' experiences during the Mexican revolution. "My mother," Nava wrote, "told us about how roving bands of soldiers would ride into town by surprise. It made no difference if they were government *federales* or revolutionaries. They both did the same things. They took food, supplies, horses, money, and kidnapped women." Nava's mother fled to the United States, as did, eventually, her future husband and brother-in-law. Nava's father "favored the revolutionary cause, but in time...[he] must have seen that fighting made little sense, because one military group was no better than the other." Because of Mexico's economic plight and its frightening revolution, over a million newcomers crossed America's southern border during the early decades of the twentieth century. With such large numbers coming to America throughout the twentieth century, "[i]t is entirely possible," historian Manuel G. Gonzales has noted, "that more Mexicans have immigrated into the United States than any other single national group, including both Germans and Italians."[26]

The newcomers from Mexico, as well as many other Spanish-speakers in the United States—unlike many of the European immigrants who were gradually shedding their ethnic identities—often clung to their cultural and linguistic heritage. Some Mexican immigrants, as Sandra Cisneros noted in her classic novel *The House on Mango Street*, were "afraid of English," but this anxiety over the American language was not the only motivation for language maintenance. Frequently living in segregated neighborhoods "[a]cross the tracks," like the Segundo Barrio in El Paso—in some cities these areas were referred to as "Little Mexico[s],"—Spanish speakers in the Southwest found security in their customs and took great pride in their mother tongue. Esquibel's Mexican father, for instance, insisted that his family speak only Spanish at home, noting to his children "'aquí en mi casa se va hablar español' (here in my house Spanish is spoken)." The children's fluency in English from public schooling caused a generational rift in the Esquibel household, and *el padre*, therefore, adamantly demanded the utilization of grammatically "correct" Spanish by his *hijos*. Perhaps reflecting his "bad encounters over language in Texas," Teodoro Esquibel became "upset" by "any childish banter...in English." For Gloria López-Stafford, the reverse occurred; she, the daughter of a Mexican immigrant and an Anglo, refused to fully learn or use English for much of her early childhood despite her father's encouragement, largely because the language "sounds ugly. And I look stupid speaking it."[27]

While *español* was an essential aspect of the immigrants' culture, so too was education. Although the prevailing view of Mexicans was that they were intellectually "inferior," "dirty and often infected with head lice" and that, as one school superintendent stated, they "care nothing about going to school," many Spanish speakers greatly desired high-quality schooling for their children. While many Mexican and "Spanish-American" parents were deeply committed to education, that commitment often hinged upon who controlled the schools and the families' economic resources. In areas where Latinos had a great deal of power in the realm of education, as in San Luis, Colorado, the "problems" associated with Spanish-speaking children, such as

high dropout rates, were significantly less pronounced. Many agricultural migrants also had a deep cultural commitment to education,
even though they did not always have the economic security to send
their children to school on a regular basis. As a migrant family in
California during the Great Depression, the Esquibels insisted that
the children receive an education, regardless of the family's financial
concerns. Esquibel remembered her mother saying: "It isn't true that
a child's only obligation is to add to the family income while she lives
at home. A person needs an education to improve her lot." Although
their parents were often devoted to education, Spanish-speaking children typically attended inferior, segregated schools, even though
children of Mexican descent were technically classified as "white"
throughout much of the Southwest. In the 1930s, prominent scholars,
such as Sanchez and Herschel T. Manuel—as well as Annie Reynolds
from the U.S. Office of Education—found that segregation, while
sometimes justified by pedagogical and linguistic considerations,
was primarily a reflection of local communities' prejudices. Citing
an official report, Manuel discovered that many Texans believed that
"[t]he American children...do not like to go to school with the dirty
'greaser' type of Mexican child.... There is but one choice in the
matter of educating these unfortunate children and that is to put the
'dirty' ones into separate schools till they learn how to 'clean up.' "[28]

Forced into segregated schools or classrooms—85 percent of
Spanish-speaking children faced segregation in 1930—Mexican
children often encountered strict English-only policies throughout
the Southwest. In Texas, for instance, some school officials suggested that students should be punished for using Spanish during
recess. In California during the 1930s, Nava's kindergarten teacher,
who "caught me speaking Spanish," "launched me across the floor
with one blow to the fanny." "Speak English, Julian," the teacher
yelled, "You are an American." But, enforcing the English-only rules
was sometimes a personal choice on the part of the educator. Nava
remembered another teacher for whom "Spanish was okay...because
she said it was a beautiful language." In spite of these monolingual
policies, bilingual education persisted in some localities. By the early
1930s, some public schools in Los Angeles began to see the utility
of emphasizing Mexican culture throughout the curriculum. "In the
regular schools," Reynolds reported in 1933, "an attempt is made to
recognize the particular needs of Mexican children and to keep activities in familiar territory." Elementary educators in Los Angeles taught
"such topics as life on a Mexican ranch, preparation of Mexican food,

[and] historical events in California under Mexico." This recognition of Mexican heritage certainly made the curriculum more relevant and accessible to many immigrant students. Completing part of her elementary schooling in California, Esquibel later complained about the Anglo-centric course of study. "Nothing," Esquibel noted about Dick, Jane, or other characters in school readers, "matched my reality."[29] Los Angeles' understanding that immigrant children required some form of bicultural instruction went a long way toward improving the school experiences of Mexican children, and this biculturalism helped lay the foundation for bilingual education in later years.

While some schools in California were experimenting with methods to meet the educational needs of Spanish-speaking children, Texas was in the forefront of the movement to instruct the Mexican newcomers and *tejanos* bilingually. Although the Lone Star State had jumped on the English-only bandwagon during the war years— passing monolingual legislation in 1918—by the late 1920s Texan officials revised this strict school law and allowed for elementary bilingual instruction in the counties that bordered with Mexico. In the early 1930s, Texas again modified its bilingual policy and permitted foreign-language instruction in elementary schools throughout the state. In 1933, the year in which the second modification was enacted, Reynolds reported that at an El Paso school that served Mexican upper-elementary and high school students a bilingual staff member was employed to counsel pupils about their educational and personal problems. In some of the Val Verde County schools, Manuel discovered in 1930, Spanish was used in the lower elementary grades to assist in the learning of English. A first-grade teacher in the county, Eloisa Barrera, noted that "sentences are given in Spanish, then in English. An explanation is made also by action or by picture." At the end of the interwar period, the Corpus Christi schools introduced Spanish in grades three through six and taught the language to around sixteen thousand pupils. In addition to the public schools, a number of private schools were established by Mexican Americans in Texas to shield their children from the discrimination in the public schools and to maintain their native language and culture to a higher degree, a move that was greatly facilitated by the U.S. Supreme Court's *Meyer* decision.[30]

The northward migration of African Americans, the subsequent loss of ethnic identity among the Europeans groups, the arrival of vast numbers of Mexicans, and the newcomers' desire for language maintenance, however, were not the only forces that were shifting the focus of bilingual education toward non-European peoples. Moreover, America's Spanish-speaking populations were not the only groups participating in this shift. The continuation of progressive thought, particularly its humanitarian side, had a tremendous influence on the bilingual activities of American Indians. After World War I, criticism of the federal government's treatment of Native Americans grew exponentially. The social activist John Collier, for instance, became alarmed by the poor handling of Native Americans and, in the early 1920s, helped form the American Indian Defense Association, which, in popular journals, publicly denounced federal Indian policies. The growing condemnation of the government's conduct toward America's original inhabitants facilitated the passing of the Indian Citizenship Act of 1924, which eliminated the Native Americans' previous status of government wards and, thus, granted citizenship to all Indians "born within the territorial limits of the United States." The gathering criticism after the war also led Secretary of the Interior Hubert Work to request the Institute for Government Research to conduct an independent examination of the government's Indian policies. The institute's survey of Indian affairs was directed by Lewis Meriam, and the 1928 document, *The Problem of Indian Administration*—often called the Meriam Report—emerged as a scathing critique of the U.S. government's practices regarding Native Americans. The progressive educational leader W. Carson Ryan, Jr. authored much of the Meriam Report's section on education and recommended that schools for American Indians emphasize Native American history, art, and culture. The curriculum, the Meriam Report emphatically stated, should be based on "Indian interests" and "Indian experiences." This pulling away from the strict Americanization and cultural extinction policies required an entirely different approach to Native American schooling. "The most fundamental need in Indian education," the 1928 report declared, "is a change in point of view."[31]

This change in point of view began to be realized over the subsequent decades, decades that witnessed a reversal of many of the federal government's Native American policies, including those related to language. The progressive-oriented view of Indian policies took hold largely because of personnel changes within the Bureau of Indian Affairs (BIA). For instance, Collier, the outspoken critic of the BIA, became the commissioner of Indian affairs in 1933. During Collier's tenure, the Wheeler-Howard Act easily passed in 1934; its first section moved through both houses of the U.S. Congress and "reached the President's desk...without the change of a word or a comma."

With this Indian Reorganization Act, Collier noted that the "train of evil consequences" of the General Allotment Act of 1887 was halted and, thus, allowed for and encouraged tribal autonomy among most Native American groups, with the notable exception of those in Oklahoma; in 1936, Congress expanded the reorganization to the Indians in the former Indian Territory with the Oklahoma Indian Welfare Act. In addition to the new land policies, the era of strict English-only programs for Indian children was coming to an end. In the midst of the Great Depression, Collier's admiration of Native American culture and communitarianism—as opposed to the individualism of a failing capitalist society—took on a new saliency and fueled a desire for biculturalism. Ryan, the progressive education advocate, was appointed director of Indian education in 1930, and he, along with his successor—the social reconstructionist Willard Beatty—infused the schools for Native Americans with the new "point of view" that insisted that education be connected to the cultural experiences of American Indian children, which naturally led to the experimentation with bilingual education. Under the influence of progressive thought, the federal government came to view Native American dual-language instruction as less threatening than it had been in the past; bilingualism, along with native arts and crafts, had become an acceptable mode of Indian expression.[32]

During Beatty's tenure as the director of Indian education (1936–1952), a number of dictionaries, grammars, and bilingual textbooks were developed for Native American pupils, including the *Indian Life Series*. Although often containing an assimilation message, the bilingual readers for Pueblo, Sioux, and Navajo students fostered the utilization of native tongues in Indian schools and validated the children's culture. Ann Nolan Clark, a longtime teacher at an Indian school in the Southwest, was instrumental in the development and writing of the *Life Series* books for the Pueblo and Sioux children; her Pueblo students even helped write one of her English-language readers which was subsequently published under the title *In My Mother's House*. Much of the native-language portions of the bilingual readers were translated from the English by Native Americans, such as Albert Yava. Yava, a Tewa, translated some of the Hopi readers, since he was proficient in that language as well as several others; for a multilingual translator such as Yava, it was—like Virginia Woolf wrote of another language expert—"as if language were wine upon his lips." These early experiments with bilingual readers and education set the stage for the language maintenance (and language restoration) programs during the post–World War II era and contributed to a dramatic reconstruction of the polyglot boardinghouse.[33]

This reconstruction of the polyglot boardinghouse during the interwar years was indeed dramatic because of the stepped-up bilingualism of non-European peoples. Although bilingual education did not "stop" after World War I, immigration restriction, loss of ethnic identification, and new approaches to Americanization facilitated a general lessening of the European immigrants' involvement in dual-language instruction. While the once prominent tenants were losing interest in the boardinghouse, its other occupants came into the parlor to showcase their linguistic and cultural endeavors. The New Mexicans, proud of their Spanish heritage, continued their public bilingual programs, even during times of intense national conformity. The continuous arrival of Mexican immigrants facilitated the endurance of Spanish-English education in many areas of the Southwest. Spanish speakers, however, were not the only non-Europeans who participated in public bilingual schooling during the interwar period. American Indians, long forced into strict English-only classrooms, encountered a new environment with regard to language maintenance. With progressive education gaining force within the Bureau of Indian Affairs, particularly the social reconstructionist strand, many Native Americans were able to use their mother tongues in government schools.

Epilogue

The Federal Landlord

The interwar years set the stage for the emergence of the modern, federally funded bilingual education movement, a movement that, unlike the earlier manifestations of bilingual schooling, has been carefully chronicled by numerous scholars. The post–World War II history of the polyglot boardinghouse has become a familiar story largely because of the exceptional work of scholars such as Theodore Andersson, Joshua Fishman, Guadalupe San Miguel, James Crawford, Rubén Donato, and Diego Castellanos.[1] The brief highlights of that story include, most notably, the passing of the Bilingual Education Act (BEA) in 1968, legislation that encouraged schools to address the needs of English-language learners and that provided federal funds for those endeavors. Two years later, Chinese Americans filed suit against the public schools in San Francisco, arguing that their children were not receiving the special help required by those unfamiliar with the American language. When the case, *Lau v. Nichols*, was decided in 1974, the U.S. Supreme Court ruled that public schools that did not address the needs of English-language learners were violating the Civil Rights Act of 1964, an act that guaranteed non-discriminatory practices in public institutions, including schools. By not providing the plaintiffs with some sort of special program, the Court noted, the San Francisco schools were not giving the non-English-speaking students an equal educational opportunity and, thus, were discriminating against them. To enforce the Court's decision, the Office of Education issued the "Lau remedies" the following year, a set of guidelines for schools to identify and provide services for English-language learners, particularly bilingual education. Although the Bilingual Education Act and the subsequent *Lau* ruling certainly encouraged bilingual instruction in America's public schools, the federal government, in fact, did not mandate a strict dual-language approach to education; as long as school programs—either bilingual or not—catered to the special needs of English-language learners they were following the letter of the law. Without a clear directive for bilingual education programs or the aims of those programs—such as transitioning children to English or maintaining mother tongues—the government has contributed to the continuing debate surrounding dual-language instruction.[2]

The movement of bilingual education into the orbit of the federal government was, of course, a dramatic shift in educational policy. As with dual-language instruction during the nineteenth and early-twentieth centuries, the nature of bilingual education during the post–World War II years was a reflection of larger social and political conditions, and there were momentous changes in the period that fostered the federal involvement in language education. World War II and the Cold War intensified an interest in the U.S. Office of Education's Foreign Languages in Elementary Schools (FLES) program, which, although beginning in the interwar era, grew dramatically in the 1950s as a means of promoting more rigorous instruction in order to gain the upper hand over the Soviet Union. When the Russians successfully launched *Sputnik* in 1957, the first manmade satellite to orbit the Earth, the United States responded with the National Defense Education Act (NDEA) the following year. The legislation stated that "an insufficient proportion of our population [is being] educated in science, mathematics, and foreign languages." The goal, therefore, was to provide more training in these critical areas in order to identify "the talent of our Nation," a sentiment echoed by Vice Admiral Hyman Rickover and James Conant; both of these Cold Warriors, in widely read works, called for using the schools to assist in the struggle against the Soviet Union. Because knowledge of foreign languages was recognized as vital to national security, more federal funds were made available for the study of languages other than English.[3]

Perhaps nothing pushed bilingual instruction in the federal government's arms more than the Civil Rights movement and President Lyndon B. Johnson's "War on Poverty." Civil Rights leaders and progressive educational reformers were committed to ending the discriminatory practices of schools and, influenced by new approaches to psychology, became deeply concerned about the self-esteem of students, a concern that was reflected in numerous popular publications, including those of the educational theorist Paul Goodman and the journalist Paul Cowan. In 1954, the U.S. Supreme Court's ruled in

Brown v. Board of Education that segregation was unconstitutional and, capturing the Civil Rights era's growing interest in the self-esteem of youngsters, noted that the officially sanctioned segregation of black students "solely because of their race generates a feeling of inferiority as to their status in the community that may affect their hearts and minds in a way unlikely ever to be undone." Race, ethnicity, and social class have been intimately interwoven in the United States, and, therefore, Civil Rights legislation also addressed the poverty that many minorities faced. In 1965, Johnson's Elementary and Secondary Education Act sought to improve the academic success of low-income students by "bring[ing] education to millions of disadvantaged youth who need it most." "As a former teacher," Johnson noted upon signing the bill, "I have great expectations of what this law will mean for all of our young people.... I believe deeply no law I have signed or will ever sign means more to the future of America."[4]

It was within this context of concern for the civil liberties of minorities of all sorts that scholars and activists turned their attention to bilingual instruction as a means of ameliorating the educational disadvantages that foreign-language-speaking students faced. Andersson, Fishman, and George Sanchez, for instance, diligently promoted bilingual education and advocated for the passage of the Bilingual Education Act. In a 1960 letter to a Texas educational official, Sanchez noted, "If 80 per cent of the non-English-speaking children are having to spend two or more years in the first grade there must be something radically wrong with the schools," not with the pupils.[5] Sanchez, a childhood friend of Johnson, contacted the new president and asked him to call a national conference on the plight of Spanish-speakers. The subsequent conference in El Paso brought the educational needs of the Mexican Americans before a national audience. In the late 1960s, Ralph Yarborough, a U.S. senator from Texas, introduced legislation that aimed at addressing those educational needs, particularly those related to language. Drawing on phraseology similar to that of the *Brown* decision, Senator Yarborough stated in 1967 that "the Mexican American child is wrongly led to believe from his first day of school that there is something wrong with him, because of his language. This misbelief soon spreads to the image he has of his culture, of the history of his people, and of his people themselves." This internalization of cultural hatred, leading to self-loathing (as with segregated African American children), did "psychological harm" to students, as Andersson stated before the U.S. House of Representatives when it was considering the Bilingual Education Act. Fishman also presented expert testimony in support of bilingual schooling, a form of schooling which, by the late 1960s, had models of excellence in the Miami area that were established by Cuban exiles; fleeing the island after the "*barbudos*, or bearded ones," came to power in January 1959, educated, middle-class Cuban refugees, believing that one day soon they would return to their homeland, maintained their language within the Dade County schools. In 1968, Johnson signed the Bilingual Education Act into law, and the legislation became Title VII of his Elementary and Secondary Education Act.[6]

Federally supported bilingual instruction is, as San Miguel has pointed out, a "contested policy," and the Bilingual Education Act certainly was not the solution to the centuries-old debate over dual-language instruction. While the *Brown* decision was a major victory for Civil Rights leaders, it did not *end* segregation, although *Brown* did end its legality. As the activist Jonathan Kozol has pointed out, "apartheid schooling" in America actually has increased in recent decades. Similarly, bilingual education, seemingly triumphant after the BEA and the *Lau* decision, has faced enormous setbacks and criticisms since the 1980s. It is likely that dual-language instruction in the nation's public schools, whether federally funded or not, will continue to be a "contested" practice because, as this book has argued, the polyglot boardinghouse is a reflection of American values and ideals, notions that are continually in flux and, therefore, perennially disputed.[7]

Notes

Introduction: Drafting the Blueprints for This Old Boardinghouse

1. L. N. Hines, ed., *Proceedings and Papers of the Indiana State Teachers' Association, October 31, and November 1, 2, 3, 1917* (n.p., n.d.), 350.
2. Paul J. Ramsey, "The War against German-American Culture: The Removal of German-Language Instruction from the Indianapolis Schools, 1917–1919," *Indiana Magazine of History* 98, no. 4 (2002): 285–303; Hines, *Proceedings, 1917*, 350.
3. Kurt Vonnegut, *God Bless You, Mr. Rosewater* (New York: Delta, 1998), 28–29.
4. Benjamin Franklin, "The German Language in Pennsylvania," in *Language Loyalties: A Source Book on the Official English Controversy*, ed. James Crawford (Chicago: University of Chicago Press, 1992), 19.
5. H. L. Mencken, *The American Language: An Inquiry into the Development of English in the United States*, Supplement I (New York: Alfred A. Knopf, 1966), 138–139.
6. John B. Peaslee, "Instruction in German and Its Helpful Influence on Common-School Education as Experienced in the Public Schools of Cincinnati" (paper presented at the National German American Teachers' Association, Chicago, IL, July 19, 1889, University of Illinois Library, Urbana, IL), 8–9.
7. Ramsey, "War against German-American Culture," 285–303; Paul J. Ramsey, "'Let Virtue Be Thy Guide, and Truth Thy Beacon-Light': Moral and Civic Transformation in Indianapolis's Public Schools," in *Civic and Moral Learning in America*, ed. Donald Warren and John J. Patrick (New York: Palgrave Macmillan, 2006), 135–151; Gustavus Ohlinger, *The German Conspiracy in American Education* (New York: George H. Doran Company, [1919?]), 108–113; Gustavus Ohlinger, *Their True Faith and Allegiance* (New York: Macmillan Company, 1916), 43.
8. See Lou Dobbs, "English-Only Advocates See Barriers to Bill Easing Up," available at www.cnn.com/2005/US/04/18/official.english/index.html; Rachel L. Swarns, "Senate Deal on Immigration Falters," *New York Times*, April 7, 2006; Rachel L. Swarns, "Immigrants Rally in Scores of Cities for Legal Status," *New York Times*, April 11, 2006; "Bilingual Material in Libraries

Draws Some Criticism," *New York Times*, September 5, 2005; Mary Ann Zehr, "English-Only Advocate Uses Ariz. State Office to Carry out Mission," *Education Week*, February 1, 2006.

9. Eugene E. García, *Teaching and Learning in Two Languages: Bilingualism and Schooling in the United States* (New York: Teachers College Press, 2005), 23–62, 77–99; Zehr, "English-Only Advocate"; Dobbs, "English-Only Advocates."

10. William J. Bennett, *The De-Valuing of America: The Fight for Our Culture and Children* (New York: Summit, 1992), 54; Diane Ravitch, "Politicization and the Schools: The Case of Bilingual Education," *Proceedings of the American Philosophical Society* 129, no. 2 (1985): 127.

11. Carlos Kevin Blanton, *The Strange Career of Bilingual Education in Texas, 1836–1981* (College Station: Texas A&M University Press, 2004), 3; Steven L. Schlossman, "Is There an American Tradition of Bilingual Education?: German in the Public Elementary Schools, 1840–1919," *American Journal of Education* 91, no. 2 (1983): 139–140.

12. For bilingual education's more recent history, see Guadalupe San Miguel, Jr., *Contested Policy: The Rise and Fall of Federal Bilingual Education in the United States, 1960–2001* (Denton, TX: University of North Texas Press, 2004); Guadalupe San Miguel, Jr., *"Let All of Them Take Heed": Mexican Americans and the Campaign for Educational Equality in Texas, 1910–1981* (Austin: University of Texas Press, 1987), 181–210; Rubén Donato, *The Other Struggle for Equal Schools: Mexican Americans during the Civil Rights Era* (Albany: State University of New York Press, 1997), 103–143; K. Tsianina Lomawaima and Teresa L. McCarty, *"To Remain an Indian": Lessons in Democracy from a Century of Native American Education* (New York: Teachers College Press, 2006), 114–133.

13. John Bodnar, *The Transplanted: A History of Immigrants in Urban America* (Bloomington: Indiana University Press, 1985), 189–197; John Higham, *Strangers in the Land: Patterns of American Nativism, 1860–1925* (New Brunswick: Rutgers University Press, 1955), 59–60, 208; Oscar Handlin, "Education and the European Immigrant, 1820–1920," in *American Education and the European Immigrant, 1840–1940*, ed. Bernard J. Weiss (Urbana: University of Ilinois Press, 1982), 12–15; David B. Tyack, *The One Best System: A History of American Urban Education* (Cambridge, MA: Harvard University Press, 1974), 106–109, 229–255.

14. Selma Cantor Berrol, *Growing Up American: Immigrant Children in America, Then and Now* (New York: Twayne Publishers, 1995), 31–59; Selma Cantor Berrol, *Immigrants at School: New York City, 1898–1914* (New York: Arno Press, 1978). There is a notable exception to this peripheral treatment of bilingual schooling, however. Francesco Cordasco has devoted a great deal of his career to the history of bilingual education, particularly by compiling bibliographies and editing collections of primary documents on the topic; see, for example, Francesco Cordasco, *Bilingual Schooling in the United States: A Sourcebook for Educational Personnel* (New York: McGraw-Hill, 1976); Francesco Cordasco, ed., *The Italian*

Community and Its Language in the United States: The Annual Reports of the Italian Teachers Association (Totowa, NJ: Rowman and Littlefield, 1975).

15. John Higham, "Changing Paradigms: The Collapse of Consensus History," *The Journal of American History* 76, no. 2 (1989): 460–466; Paul J. Ramsey, "Histories Taking Root: The Contexts and Patterns of Educational Historiography during the Twentieth Century," *American Educational History Journal* 34, no. 2 (2007): 350–354. John Bodnar, "Schooling and the Slavic-American Family, 1900–1940," in *American Education and the European Immigrant*, 78–95; Schlossman, "Is There an American Tradition of Bilingual Education?," 139–186; LaVern J. Rippley, *The German-Americans* (Boston: Twayne Publishers, 1976), 120–128; Frederick C. Luebke, *Bonds of Loyalty: German-Americans and World War I* (DeKalb, IL: Northern Illinois University Press,1974); Frederick C. Luebke, "The German-American Alliance in Nebraska, 1910–1917," *Nebraska History* 49, no. 2 (1968): 177–179; John M. Nieto-Phillips, *The Language of Blood: The Making of Spanish-American Identity in New Mexico, 1880s–1930s* (Albuquerque: University of New Mexico Press, 2004), 197–205; Manuel G. Gonzales, *Mexicanos: A History of Mexicans in the United States* (Bloomington: Indiana University Press, 2000), 234–237; Jon Reyhner and Jeanne Eder, *American Indian Education: A History* (Norman: University of Oklahoma Press, 2004), 78–80. There are, of course, plenty of historians who do not limit their scholarship to one ethnic group; classic examples are Bodnar's *Transplanted* and Ronald Takaki's *A Different Mirror: A History of Multicultural America* (Boston: Little, Brown and Company, 1993).

16. Frederick C. Luebke, "Turnerism, Social History, and the Historiography of European Ethnic Groups in the United States," in *Germans in the New World: Essays in the History of Immigration* (Urbana: University of Illinois Press, 1990), 138. For an example of celebratory history, see Józef Miaso, *The History of the Education of Polish Immigrants in the United States*, trans. Ludwik Krzyżanowski (New York: Kosciuszko Foundation, 1977).

17. Russell A. Kazal, *Becoming Old Stock: The Paradox of German-American Identity* (Princeton: Princeton University Press, 2004), 5.

18. Einar Haugen, *The Norwegian Language in America: A Study of Bilingual Behavior*, 2 vols. (Philadelphia: University of Pennsylvania Press, 1953); Shirley Brice Heath, "A National Language Academy?: Debate in the New Nation," *International Journal of the Sociology of Language* 11 (1976): 9–43; Theodore Andersson and Mildred Boyer, *Bilingual Schooling in the United States: History, Rationale, Implications, and Planning*, vol. 1 (Austin: Southwest Educational Development Laboratory, 1970); Joshua A. Fishman, *Language Loyalty in the United States: The Maintenance and Perpetuation of Non-English Mother Tongues by American Ethnic and Religious Groups* (London: Mouton and Co., 1966); Blanton, *Strange Career*, 121–136.

19. Terrence G. Wiley, "Heinz Kloss Revisited: National Socialist Ideologue or Champion of Language-Minority Rights?," *International Journal of*

the Sociology of Language 154 (2002): 83–97; Francesco Cordasco, ed., *Dictionary of American Immigrant History* (Metuchen, NJ: Scarecrow Press, 1990), 468–469; Heinz Kloss, *The American Bilingual Tradition* (1977; reprint, with a new introduction by Reynaldo F. Macias and Terrence G. Wiley, McHenry, IL: Center for Applied Linguistics and Delta Systems Co., 1998); Heinz Kloss, *Das Nationalitätenrecht derVereinigten Staaten von Amerika* (Wien: Wilhelm Braumüller, 1963); Heinz Kloss, ed., *Laws and Legal Documents Relating to Problems of Bilingual Education in the United States* (Washington, DC: Center for Applied Linguistics, 1971).

20. Blanton, *Strange Career*, 4; William S. Burroughs, *My Education: A Book of Dreams* (New York: Penguin, 1995), 2; Reynaldo F. Macías and Terrence G. Wiley, "Introduction to the Second Edition," in *The American Bilingual Tradition*, Heinz Kloss (McHenry, IL: Center for Applied Linguistics and Delta Systems, 1998), xii–xiii.

21. Dobbs, "English-Only Advocates."

22. Rogers M. Smith, *Civic Ideals: Conflicting Visions of Citizenship in U.S. History* (New Haven: Yale University Press, 1997), 8–26; Barry L. Bull, Royal T. Fruehling, and Virgie Chatterg, *The Ethics of Multicultural and Bilingual Education* (New York: Teachers College Press, 1992), 24–29.

23. Smith, *Civic Ideals*, 13–39; E. P. Thompson, *The Making of the English Working Class* (New York: Vintage Books, 1966), 11.

24. Jonathan Zimmerman, "Ethnics against Ethnicity: European Immigrants and Foreign-Language Instruction, 1890–1940," *Journal of American History* 88, no. 4 (2002): 1383–1404; Thompson, *The Making of the English Working Class*, 194–207.

25. Michael R. Olneck and Marvin Lazerson, "The School Achievement of Immigrant Children: 1900–1930," *History of Education Quarterly* 14, no. 4 (1974): 453–482; David Hogan, "Education and the Making of the Chicago Working Class," *History of Education Quarterly* 18, no. 3 (1978): 227–270; Joel Perlmann, *Ethnic Differences: Schooling and Social Structure among the Irish, Italians, Jews, and Blacks in an American City, 1880–1935* (Cambridge: Cambridge University Press, 1988), 1–12, 83–121; Ramsey, "War against German-American Culture," 285–303; Nieto-Phillips, *The Language of Blood*, 73–80.

26. Michel Foucault, *The Archaeology of Knowledge and the Discourse on Language*, trans. A. M. Sheridan Smith (New York: Pantheon Books, 1972), 216, 219.

27. John McWhorter, *The Power of Babel: A Natural History of Language* (New York: Perennial, 2003), 63–74; Ramsey, "Let Virtue Be Thy Guide," 135–151; Zimmerman, "Ethnics against Ethnicity," 1383–1404.

28. Donna Christian and Fred Genesee, *Bilingual Education* (Alexandria, VA: TESOL, 2001), 1–3; Haugen, *The Norwegian Language in America*, vol. 1, 7.

29. B. Edward McClellan, *Moral Education in America: Schools and the Shaping of Character from Colonial Times to the Present* (New York: Teachers College Press, 1999), xii.

1   Laying the Foundation for the Boardinghouse: The Context of Nineteenth-Century Schooling and Bilingualism

1. Thomas Jefferson, "Inaugural Address," in *Our Nation's Archive: The History of the United States in Documents*, ed. Erik Bruun and Jay Crosby (New York: Tess Press, 1999), 227–228; Charles A. Beard, *Economic Origins of Jeffersonian Democracy* (1915; reprint, New York: Free Press, 1965), 421–428, 450–453; Ronald Takaki, *A Different Mirror: A History of Multicultural America* (Boston: Little, Brown and Company, 1993), 166; Rogers M. Smith, *Civic Ideals: Conflicting Visions of Citizenship in U.S. History* (New Haven: Yale University Press, 1997), 165–196; Joseph J. Ellis, *Founding Brothers: The Revolutionary Generation* (New York: Vintage Books, 2002), 15–16.

2. Takaki, *A Different Mirror*, 139–148, 192–198; John Bodnar, *The Transplanted: A History of Immigrants in Urban America* (Bloomington: Indiana University Press, 1985), 1–30; U.S. Immigration Commission, *Statistical Review of Immigration, 1820–1910—Distribution of Immigrants, 1850–1900* (Washington, DC: Government Printing Office, 1911), 4–24; "Roosevelt Demands Speed-Up of War," *New York Times*, August 27, 1918, 8.

3. Thomas Jefferson, "A Bill for the More General Diffusion of Knowledge, 1779," in *The School in the United States: A Documentary History*, ed. James W. Fraser (New York: McGraw-Hill, 2001), 19–24; Thomas Jefferson, "Notes on the State of Virginia, 1783," in *The School in the United States*, 24–27; Benjamin Rush, "Thoughts upon the Mode of Education Proper in a Republic, 1786," in *The School in the United States*, 28.

4. Andrew Jackson, "Check This Spirit of Monopoly," in *Our Nation's Archive*, 294; Arthur M. Schlesinger, Jr., *The Age of Jackson* (Boston: Little, Brown and Company, 1946), 36–44, 74–87, 210–226; James Oakes, "The Ages of Jackson and the Rise of American Democracies," *The Journal of the Historical Society* 6, no. 4 (2006): 491–500; Sean Wilentz, "Politics, Irony, and the Rise of American Democracy," *The Journal of the Historical Society* 6, no. 4 (2006): 537–539; Smith, *Civic Ideals*, 170–173, 213–216; Walter Dean Burnham, "Table I: Summary: Presidential Elections, USA, 1788–2004," *The Journal of the Historical Society* 7, no. 4 (2007): 530–537; David Nasaw, *Schooled to Order: A Social History of Public Schooling in the United States* (Oxford: Oxford University Press, 1981), 39–43; Catharine E. Beecher, "An Essay on the Education of Female Teachers for the United States, 1835," in *The School in the United States*, 62; Kathryn Kish Sklar, *Catharine Beecher: A Study in American Domesticity* (New York: W. W. Norton and Company, 1976), 122–137.

5. Henry Boernstein, *The Mysteries of St. Louis*, trans. Friedrich Münch (1851; modern edition by Steven Rowan and Elizabeth Sims, Chicago: Charles H. Kerr Publishing Company, 1990), 24; Oakes, "The Ages of Jackson," 491–500; Wilentz, "Politics, Irony, and the Rise of American Democracy," 537–539.

6. B. Edward McClellan, *Moral Education in America: Schools and the Shaping of Character from Colonial Times to the Present* (New York: Teachers College Press, 1999), 9–14; Gordon S. Wood, *The Radicalism of the American Revolution* (New York: Vintage Books, 1993), 43–77; Lawrence A. Cremin, *American Education: The Colonial Experience, 1607–1783* (New York: Harper and Row, 1970), 480–491; Iain Pears, *An Instance of the Fingerpost* (New York: Riverhead Books, 2000), 328; Robert H. Wiebe, *The Segmented Society: An Introduction to the Meaning of America* (New York: Oxford University Press, 1975), 30–31; Philip Greven, *The Protestant Temperament: Patterns of Child-Rearing, Religious Experience, and the Self in Early America* (Chicago: University of Chicago Press, 1988), 206–217, 358–361; Thomas S. Kidd, "What Happened to the Puritans?," *Historically Speaking* 7, no. 1 (2005): 32–34.

7. McClellan, *Moral Education in America*, 9–14; Barnard Bailyn, *Education and the Forming of American Society* (New York: W. W. Norton and Company, 1972), 10–11; Carl F. Kaestle, *Pillars of the Republic: Common Schools and American Society, 1780–1860* (New York: Hill and Wang, 2001), 30–61; William J. Reese, *The Origins of the American High School* (New Haven: Yale University Press, 1995), 1–28.

8. Kaestle, *Pillars of the Republic*, 63–66; Schlesinger, *Age of Jackson*, 144–158, 177–189, 334–349; Michael B. Katz, *The Irony of Early School Reform: Educational Innovation in Mid-Nineteenth Century Massachusetts* (New York: Teachers College Press, 2001), 5–11, 222–223; Eric Hobsbawm, *The Age of Capital, 1848–1875* (New York: Vintage Books, 1996), 309–311.

9. Kaestle, *Pillars of the Republic*, 63–67; Katz, *Irony*, 222; Takaki, *A Different Mirror*, 139–160; Thomas Bender, *Toward an Urban Vision: Ideas and Institutions in Nineteenth-Century America* (Lexington: University Press of Kentucky, 1975), 3–17; U.S. Immigration Commission, *Statistical Review of Immigration*, 4–5, 8–11; Bodnar, *Transplanted*, 64.

10. Wiebe, *The Segmented Society*, 31; Schlesinger, *Age of Jackson*, 334–336; Horace Mann, "Tenth and Twelfth Annual Reports to the Massachusetts Board of Education," in *The School in the United States*, 54. Italics added.

11. Merle Curti, *The Social Ideas of American Educators*, revised ed. (Paterson, NJ: Littlefield, Adams and Co., 1959), 101–168; James W. Fraser, *Between Church and State: Religion and Public Education in a Multicultural America* (New York: St. Martin's Griffin, 2000), 23–47; Kaestle, *Pillars of the Republic*, 75–103; Katz, *Irony*, 27–50, 159–160.

12. Wiebe, *The Segmented Society*, 31; Alexis de Tocqueville, *Democracy in America and Two Essays on America*, ed. Isaac Kramnick, trans. Gerald E. Bevan (New York: Penguin Books, 2003), 299, 301; Marcus Aurelius, *Meditations*, trans. Maxwell Staniforth (Baltimore: Penguin Books, 1964), 104.

13. Wiebe, *The Segmented Society*, 31.

14. William J. Reese, *America's Public Schools: From the Common School to "No Child Left Behind"* (Baltimore: Johns Hopkins University Press, 2005), 28–32; McClellan, *Moral Education in America*, 26; *By-Laws of the School Officers and Trustees of the Nineteenth Ward, and Rules for the Government*

of Schools (New York: J. Youdale, 1863), 4, 27, 30; Marcius Willson, *The Fourth Reader of the United States Series* (New York: Harper and Brothers, 1872), 21–22, 55–57, 117–118; Benjamin B. Comegys, ed., *A Primer of Ethics* (Boston: Ginn and Company, 1891), 48, 124; Asa Fitz, *The American School Hymn Book* (Boston: Crosby, Nichols and Company, 1854), 94; Ruth Miller Elson, *Guardians of Tradition: American Schoolbooks of the Nineteenth Century* (Lincoln: University of Nebraska Press, 1964), 41–48, 226.

15. Common Schools of Pennsylvania, *Report of the Superintendent of Common Schools of the Commonwealth of Pennsylvania, for the Year Ending June 3, 1867* (Harrisburg: Singerly and Myers, 1868), xli; Mark Twain, *Life on the Mississippi* (Mineola, NY: Dover Publications, 2000), 51; italics in the original.

16. Kaestle, *Pillars of the Republic*, 30–74; Katz, *Irony*, 224; Reese, *America's Public Schools*, 26, 43–44; John L. Rury, "Social Capital and the Common Schools," in *Civic and Moral Learning in America*, ed. Donald Warren and John J. Patrick (New York: Palgrave Macmillan, 2006), 77–80; Tocqueville, *Democracy in America*, 233.

17. Beecher, "An Essay on the Education of Female Teachers," 62; Katz, *Irony*, 19–62, 80–93, 115–160; Maris A. Vinovskis, *The Origins of Public High Schools: A Reexamination of the Beverly High School Controversy* (Madison: University of Wisconsin Press, 1985), 79–82, 104–108; Fraser, *Between Church and State*, 49–65; Lloyd P. Jorgenson, *The State and the Non-Public School, 1825–1925* (Columbia: University of Missouri Press, 1987), 57–85; Ellwood P. Cubberley, *Public Education in the United States: A Study and Interpretation of American Educational History* (Boston: Houghton Mifflin Company, 1919), 115–160; Nasaw, *Schooled to Order*, 29–84.

18. Gordon S. Wood, "The Localization of Authority in the 17th-Century English Colonies," *Historically Speaking* 8, no. 6 (2007): 2–5; Wiebe, *The Segmented Society*, 29, 35, 55; David B. Tyack, *The One Best System: A History of American Urban Education* (Cambridge, MA: Harvard University Press, 1974), 13–21, 28–77; Reese, *America's Public Schools*, 28–29; Kaestle, *Pillars of the Republic*, 147–149; Rury, "Social Capital and the Common Schools," 72.

19. Shirley Brice Heath, "Why No Official Tongue?," in *Language Loyalties: A Source Book on the Official English Controversy*, ed. James Crawford (Chicago: University of Chicago Press, 1992), 20–31; Heinz Kloss, *The American Bilingual Tradition* (1977; reprint, with a new introduction by Reynaldo F. Macias and Terrence G. Wiley, McHenry, IL: Center for Applied Linguistics and Delta Systems Co., 1998), 10; Benedict Anderson, *Imagined Communities: Reflections on the Origin and Spread of Nationalism*, revised ed. (New York: Verso, 1991), 47–65.

20. Heath, "Why No Official Tongue?," 30; Anderson, *Imagined Communities*, 47, 67–81; Noah Webster, "On the Education of Youth in America, 1790," in *The School in the United States*, 36; Kaestle, *Pillars of the Republic*, 99; Cremin, *American Education: The Colonial Experience*, 568–570; John Russell Bartlett, *Dictionary of Americanisms: A Glossary of Words and*

Phrases, Usually Regarded as Peculiar to the United States (1848; reprint, with a new introduction by Richard Lederer, Hoboken, NJ: John Wiley and Sons, Inc., 2003), vi–vii; H. L. Mencken, *The American Language: An Inquiry into the Development of English in the United States*, 4th ed. (New York: Alfred A. Knopf, 1936), 23, 90–100, 134.

21. *Atlas of American History* (Skokie, IL: Houghton Mifflin / Rand McNally, 1993), 27; Wiebe, *The Segmented Society*, 29.

22. Francis A. Walker, *The Statistics of the Population of the United States, Embracing the Tables of Race, Nationality, Sex, Selected Ages, and Occupations*, ninth census, vol. 1 (Washington, DC: Government Printing Office, 1872), xvii; Takaki, *A Different Mirror*, 84–98.

23. Bodnar, *Transplanted*, 6–16; Peter Winkel, "Mid-19th Century Emigration from Hesse-Darmstadt to Trenton, New Jersey," in *Emigration and Settlement Patterns of German Communities in North America*, ed. Eberhard Reichmann, LaVern J. Rippley, and Jörg Nagler (Indianapolis: Max Kade German-American Center, 1995), 211–222; Einar Haugen, *The Norwegian Language in America: A Study of Bilingual Behavior*, vol. 1 (Philadelphia: University of Pennsylvania Press, 1953), 18–28.

24. Bodnar, *Transplanted*, 6–16; Haugen, *The Norwegian Language in America*, vol. 1, 18–28; Odd S. Lovoll, *The Promise of America: A History of the Norwegian-American People* (Minneapolis: University of Minnesota Press, 1984), 23–29; Winkel, "Mid-19th Century Emigration," 212; Henry Boernstein, *Memoirs of a Nobody: The Missouri Years of an Austrian Radical, 1849–1866*, trans. and ed. Steven Rowan (St. Louis: Missouri Historical Society, 1997), 42–67; Carolyn R. Toth, *German-English Bilingual Schools in America: The Cincinnati Tradition in Historical Context* (New York: Peter Lang, 1990), 43–44; Walter D. Kamphoefner, Wolfgang Helbich, and Ulrike Sommer, eds., *News from the Land of Freedom: German Immigrants Write Home*, trans. Susan Carter Vogel (Ithaca: Cornell University Press, 1991), 149–152, 299–301; Johann Bauer, June 10, 1855, in *News from the Land of Freedom*, 155.

25. Selma Cantor Berrol, *Growing Up American: Immigrant Children in America, Then and Now* (New York: Twayne Publishers, 1995), 2–14; Herman Melville, *The Confidence Man* (New York: Prometheus Books, 1995), 25–26; Johann Bauer, May 11, 1854, in *News from the Land of Freedom*, 153; Wilhelmina Stille, [October 1837], in *News from the Land of Freedom*, 72.

26. Howard Fast, *The Immigrants* (Boston: Houghton Mifflin Company, 1977), 2; Berrol, *Growing Up American*, 2–14; Nikolaus Heck, July 1, 1854, in *News from the Land of Freedom*, 373; "New York State Commissioners of Emigration," in *Immigration as a Factor in American History*, ed. Oscar Handlin (Englewood Cliffs, NJ: Prentice-Hall, Inc., 1959), 34–38.

27. Kamphoefner et al., eds., *News from the Land of Freedom*, 299–303, 420–425; Haugen, *The Norwegian Language in America*, vol. 1, 18–27; Günter Moltmann, "'When People Migrate, They Carry Their Selves Along': Emigration and Settlement Patterns of German Communities in North America," in *Emigration and Settlement Patterns of German Communities*

in North America, xxix–xxx; Arlow W. Andersen, *Rough Road to Glory: The Norwegian-American Press Speaks Out on Public Affairs, 1875–1925* (Philadelphia: Balch Institute, 1990), 15; Carl Berthold, July 17, 1854, in *News from the Land of Freedom*, 331; Carl Berthold, August 9, 1857, in *News from the Land of Freedom*, 333.

28. Carl Witte, *Refugees of Revolution: The German Forty-Eighters in America* (Philadelphia: University of Pennsylvania Press, 1952), 122–129; Eric Hobsbawm, *The Age of Revolution, 1789–1848* (New York: Vintage Books, 1996), 234–252; George S. Clark, "The German Presence in New Jersey," in *The New Jersey Ethnic Experience*, ed. Barbara Cunningham (Union City, NJ: Wm. H. Wise and Co., 1977), 226; Kurt Vonnegut, *Palm Sunday: An Autobiographical Collage* (New York: Delta, 1999), 22–26; Dieter Cunz, *The Maryland Germans, a History* (Princeton: Princeton University Press, 1948), 248–251; Jörg Nagler, "Frontier Socialism: The Founding of New Ulm, Minnesota, by German Workers and Freethinkers," in *Emigration and Settlement Patterns of German Communities in North America*, 178–189; Boernstein, *Memoirs of a Nobody*, 176–182, 203–210, 220–236.

29. Andreas Hjerpeland to Ivar Kleiven, May 20, 1873, in *In Their Own Words: Letters from Norwegian Immigrants*, ed. and trans. Solveig Zempel (Minneapolis: University of Minnesota Press, 1991), 6; Johann C. W. Pritzlaff, April 23, 1842, in *News from the Land of Freedom*, 306; Boernstein, *Mysteries of St. Louis*, 9; Fast, *The Immigrants*, 7.

30. O. E. Rolvaag, *Giants in the Earth* (New York: Harper and Brothers Publishers, 1929), 39; Bergljot Anker Nilssen, June 4, 1923, in *In Their Own Words*, 158.

31. Carl Berthold, October 13, 1852, in *News from the Land of Freedom*, 323; Heinrich Möller, [1866?], in *News from the Land of Freedom*, 211; Heinrich Möller, January 24, 1869, in *News from the Land of Freedom*, 213.

32. Robert E. Park, *The Immigrant Press and Its Control* (1922; reprint, Westport, CT: Greenwood Press,1970), 5; Rolvaag, *Giants in the Earth*, 129, 316; Angela Heck, July 1, 1854, in *News from the Land of Freedom*, 372–373.

33. Moltmann, "When People Migrate, They Carry Their Selves Along," xvii–xxxii; Kamphoefner et al., *News from the Land of Freedom*, 206; Heinrich Möller, February 18, 1886, in *News from the Land of Freedom*, 218; Milan Kundera, *Ignorance*, trans. Linda Asher (New York: Perennial, 2002), 33.

34. Park, *The Immigrant Press*, 50–59, 105–108; Kloss, *The American Bilingual Tradition*, 133–137, 217–220; Andersen, *Rough Road to Glory*, 216–226; Carl Wittke, *The German-Language Press in America* (University of Kentucky Press, 1957), 13–15, 75–102; Louis P. Hennighausen, "Reminiscences of the Political Life of the German Americans in Baltimore during the Years 1850–1860," *Eleventh and Twelfth Annual Reports of the Society for the History of the Germans in Maryland, 1897–1898* (n.p., n.d.), 13; Boernstein, *Memoirs of a Nobody*, 118–119, 228; James P. Ziegler, *The German-Language Press in Indiana: A Bibliography* (Indianapolis: Max Kade German-American Center and Indiana German Heritage Society, Inc., 1994).

35. Boernstein, *Memoirs of a Nobody*, 117, 229–246; Nagler, "Frontier Socialism," 180–181; Martin Weitz, July 29, 1855, in *News from the Land of Freedom*, 346; Martin Weitz, August 23, 1857, in *News from the Land of Freedom*, 358; Tocqueville, *Democracy in America*, 596.

36. John T. McGreevy, *Catholicism and American Freedom: A History* (New York: W. W. Norton and Company, 2003), 30–42, 118–122; Kloss, *The American Bilingual Tradition*, 74–93; Kamphoefner et al., *News from the Land of Freedom*, 26; Olaf Morgan Norlie, *History of the Norwegian People* (Minneapolis: Augsburg Publishing House, 1925), 203; Wolfgang Grams, "The North German Lutheran Church in Cincinnati: An 'Osnabrück' Congregation (1838)," in *Emigration and Settlement Patterns of German Communities in North America*, 79–90; Bettina Goldberg, "Cultural Change in Milwaukee's German Evangelical Lutheran Congregations of the Missouri Synod, 1850–1930," in *Emigration and Settlement Patterns of German Communities in North America*, 115–128.

37. Heinz Kloss, "German-American Language Maintenance Efforts," in *Language Loyalties in the United States: The Maintenance and Perpetuation of Non-English Mother Tongues by American Ethnic and Religious Groups*, ed. Joshua A. Fishman (London: Mouton and Co., 1966), 224–229; Paul J. Ramsey, "'Let Virtue Be Thy Guide, and Truth Thy Beacon-Light': Moral and Civic Transformation in Indianapolis's Public Schools," in *Civic and Moral Learning in America*, 138–140; Russell A. Kazal, *Becoming Old Stock: The Paradox of German-American Identity* (Princeton: Princeton University Press, 2004), 7, 18; Christian Lenz, June 21, 1858, in *News from the Land of Freedom*, 135; Johann Witten, February 23, 1912, in *News from the Land of Freedom*, 274.

38. Clemens Vonnegut, *A Proposed Guide for Instruction in Morals from the Standpoint of a Freethinker, for Adult Persons Offered by a Dilettante* (1900; reprint, n.p., 1987), 6, 10–11; Hennighausen, "Reminiscences of the Political Life of the German-Americans," 4–6; Boernstein, *Mysteries of St. Louis*, 96–97; Ludwig Dilger, March 23, [1913], in *News from the Land of Freedom*, 502.

39. Lovoll, *The Promise of America*, 54–55; Norlie, *History of the Norwegian People*, 202–203; Grams, "The North German Lutheran Church," 79–90; Goldberg, "Cultural Change in Milwaukee's German Evangelical Lutheran Congregations," 115–128; McGreevy, *Catholicism and American Freedom*, 19–42; Jorgenson, *The State and the Non-Public School*, 69–108; McClellan, *Moral Education in America*, 35–41; Boernstein, *Mysteries of St. Louis*, 138, 141; Carl Blümner, March 18, 1841, in *News from the Land of Freedom*, 105–106; Ludwig Dilger, September 26, 1926, in *News from the Land of Freedom*, 509.

40. Bodnar, *Transplanted*, 118–120; Johann Witten, February 18, 1904, in *News from the Land of Freedom*, 268; Ludwig Dilger, January 30, 1909, in *News from the Land of Freedom*, 502.

41. Rolvaag, *Giants in the Earth*, 27, 61–62; Bodnar, *Transplanted*, 118–120; Johann Witten, February 23, 1912, in *News from the Land of Freedom*, 274; Ludwig Dilger, November 5, 1886, in *News from the Land of Freedom*,

494; Hennighausen, "Reminiscences of the Political Life of the German-Americans," 3–8; Boernstein, *Memoirs of a Nobody*, 220–221.

42. Anderson, *Imagined Communities*, 67–111; Kazal, *Becoming Old Stock*, 1–4; John McWhorter, *The Power of Babel: A Natural History of Language* (New York: Perennial, 2003), 53–92.

43. Jonathan Zimmerman, "Ethnics against Ethnicity: European Immigrants and Foreign-Language Instruction, 1890–1940," *Journal of American History* 88, no. 4 (2002): 1383–1404; Haugen, *The Norwegian Language in America*, vol. 1, 103–105, 154–156; Einar Haugen, *The Norwegian Language in America: A Study of Bilingual Behavior*, vol. 2 (Philadelphia: University of Pennsylvania Press, 1953), 337; Andreas Hjerpeland to Ivar Kleiven, June 23, 1874, in *In Their Own Words*, 8.

44. Grams, "The North German Lutheran Church in Cincinnati," 79–90; Michael Probstfeld, December 13, 1890, in *News from the Land of Freedom*, 238; Boernstein, *Memoirs of a Nobody*, 221, 227–228; Wittke, *The German-Language Press*, 75–102; Giles R. Hoyt, "Germans," in *Peopling Indiana: The Ethnic Experience*, ed. Robert M. Taylor and Connie A. McBirney (Indianapolis: Indiana Historical Society, 1996), 146–181; Kloss, "German-American Language Maintenance Efforts," 231–232.

45. Kurt Vonnegut, *A Man without a Country* (New York: Seven Stories Press, 2005), 51–52.

2   Building the Polyglot Boardinghouse in the Northeast and the South

1. Bergljot Anker Nilssen, January 15, 1924, in *In Their Own Words: Letters from Norwegian Immigrants*, ed. and trans. Solveig Zempel (Minneapolis: University of Minnesota Press, 1991), 164; Wilhelm Krumme, March 21, 1850, in *News from the Land of Freedom: German Immigrants Write Home*, ed. Walter D. Kamphoefner, Wolfgang Helbich, and Ulrike Sommer, trans. Susan Carter Vogel (Ithaca: Cornell University Press, 1991), 91; Adolf Douai and John Straubenmueller, "German Schools in the United States," *The American Journal of Education* 3 (1870): 581; Henry Boernstein, *Memoirs of a Nobody: The Missouri Years of an Austrian Radical, 1849–1866*, trans. and ed. Steven Rowan (St. Louis: Missouri Historical Society, 1997), 228; Johann Witten, February 3, 1908, in *News from the Land of Freedom*, 271; Theodore Stempfel, *Fifty Years of Unrelenting German Aspirations in Indianapolis* (1898; reprint, edited by Giles R. Hoyt, Claudia Grossmann, Elfrieda Lang, and Eberhard Reichmann, Indianapolis: German-American Center and Indiana German Heritage Society, Inc., 1991), 28.

2. Joseph W. Newman, "Antebellum School Reform in the Port Cities of the Deep South," in *Southern Cities, Southern Schools: Public Education in the Urban South*, ed. David N. Plank and Rick Ginsberg (New York: Greenwood Press, 1990), 17–35; Carl F. Kaestle, *Pillars of the Republic: Common Schools and American Society, 1780–1860* (New York: Hill and Wang, 2001), 75–103; Michael B. Katz, *The Irony of Early School*

Reform: Educational Innovation in Mid-Nineteenth Century Massachusetts (New York: Teachers College Press, 2001), 27–50; Heinz Kloss, "German-American Language Maintenance Efforts," in *Language Loyalties in the United States: The Maintenance and Perpetuation of Non-English Mother Tongues by American Ethnic and Religious Groups*, ed. Joshua A. Fishman (London: Mouton and Co., 1966), 226.

3. Heinz Kloss, *The American Bilingual Tradition* (1977; reprint, with a new introduction by Reynaldo F. Macias and Terrence G. Wiley, McHenry, IL: Center for Applied Linguistics and Delta Systems Co., 1998), 9; Kloss, "German-American Language Maintenance Efforts," 213.

4. Ruth Miller Elson, *Guardians of Tradition: American Schoolbooks of the Nineteenth Century* (Lincoln: University of Nebraska Press, 1964), 143–146; W. T. Harris, "German Instruction in American Schools and the National Idiosyncrasies of the Anglo-Saxons and the Germans" (paper presented at the National German-American Teachers' Association, Cleveland, OH, July 16, 1890, University of Illinois Library, Urbana, IL), 4–6; William J. Reese, "The Philosopher-King of St. Louis," in *Curriculum and Consequence: Herbert M. Kliebard and the Promise of Schooling*, ed. Barry M. Franklin (New York: Teachers College Press, 2000), 155–177; Frederick Rudolph, *The American College and University: A History* (1962; reprint, with an introduction by John R. Thelin, Athens: University of Georgia Press, 1990), 116–135, 264–286; Charles Hart Handschin, *The Teaching of Modern Languages in the United States* (Washington, DC: Government Printing Office, 1913), 33, 70; Paul J. Ramsey, "Building a 'Real' University in the Woodlands of Indiana: The Jordan Administration, 1885–1891," *American Educational History Journal* 31, no. 1 (2004): 20–28; Calvin E. Stowe, "Report on Elementary Public Instruction in Europe, 1837," in *The School in the United States: A Documentary History*, ed. James W. Fraser (Boston: McGraw-Hill, 2001), 94–98.

5. Kurt Vonnegut, *A Man without a Country* (New York: Seven Stories Press, 2005), 51; Boernstein, *Memoirs of a Nobody*, 131–132, 220–246; Carl Wittke, *The German-Language Press in America* (University of Kentucky Press, 1957), 75–102; Giles R. Hoyt, "Germans," in *Peopling Indiana: The Ethnic Experience*, ed. Robert M. Taylor and Connie A. McBirney (Indianapolis: Indiana Historical Society, 1996), 147; Kloss, "German-American Language Maintenance Efforts," 224–225.

6. Selma Cantor Berrol, *Growing Up American: Immigrant Children in America, Then and Now* (New York Twayne Publishers, 1995), 5–11; Duncan Clark, *A New World: The History of Immigration into the United States* (San Diego: Thunder Bay Press, 2000), 15–17; Kaestle, *Pillars of the Republic*, 30–61, 75–103; Henry Barnard, ed., "Subjects and Course of Instruction in City Public Schools," *The American Journal of Education* 3 (1870): 509–510; *By-Laws of the School Officers and Trustees of the Nineteenth Ward, and Rules for the Government of Schools* (New York: J. Youdale, 1863), 26–36; *A Manual of Discipline and Instruction for the Use of the Teachers of the Primary and Grammar Schools under the Charge of the Department of Public Instruction of the City of New York* (New York: Evening Post Steam Presses, 1873), 3–7.

7. Kloss, *The American Bilingual Tradition*, 114; Barnard, "Subjects and Courses of Instruction in City Public Schools," 514; L. Viereck, "German Instruction in American Schools," in *Annual Reports of the Department of the Interior for the Fiscal Year Ended June 30, 1901: Report of the Commissioner of Education*, vol. 1 (Washington, DC: Government Printing Office, 1902), 641; Douai and Straubenmueller, "German Schools in the United States," 581–586; *Sixteenth Annual Report of the Superintendent of Public Instruction of the State of New York* (Albany: Argus Company, 1870), 84–85.

8. Viereck, "German Instruction in American Schools," 641; *Twenty-Third Annual Report of the Superintendent of Public Instruction of the State of New York* (Jerome B. Parmenter, State Printer, 1877), 48–49, 363–364; Kloss, *The American Bilingual Tradition*, 114; Barnard, "Subjects and Courses of Instruction in City Public Schools," 511–517.

9. *Annual Report, State of New York, 1877*, 130–145, 237–265, 363–364; *Third Annual Report of the City Superintendent of Schools for the Year Ending July 31, 1901* (New York: J. W. Pratt Company, n.d.), 89–90, 123–127; Indianapolis Public Schools, *Annual Report of the Secretary, Business Director, Superintendent of Schools and the Librarian, 1908–1909* (n.p., n.d.), 153; Hugo Schmidt, "A Historical Survey of the Teaching of German in America," in *The Teaching of German: Problems and Methods*, ed. Eberhard Reichmann (Philadelphia: National Carl Schurz Association, 1970), 3–7; Handschin, *The Teaching of Modern Languages*, 28–29, 94–100; Edwin H. Zeydel, "The Teaching of German in the United States from the Colonial Times to the Present," *The German Quarterly* 37 (1964): 358–360; Viereck, "German Instruction in American Schools," 641; Kloss, *The American Bilingual Tradition*, 97, 217–219.

10. David A. Gerber, "Language Maintenance, Ethnic Group Formation, and the Public Schools: Changing Patterns of German Concern, Buffalo, 1837–1874," *Journal of American Ethnic History* 4, no. 1 (1984): 31–61; Paul Rudolph Fessler, "Speaking in Tongues: German-Americans and the Heritage of Bilingual Education in American Public Schools" (PhD diss, Texas A&M University, 1997), 35–39.

11. Viereck, "German Instruction in American Schools," 640–643; Handschin, *The Teaching of Modern Languages*, 72; *Annual Report, New York City, 1901*, 89–90; Selma Cantor Berrol, *Immigrants at School: New York City, 1898–1914* (New York: Arno Press, 1978), 3; *Report of the Secretary of the Interior, Being Part of the Message and Documents Communicated to the Two Houses of Congress at the Beginning of the Third Session of the Forty-First Congress*, vol. 2 (Washington, DC: Government Printing Office, 1870), 344–352; *Report of the Secretary of the Interior, Being Part of the Message and Documents Communicated to the Two Houses of Congress at the Beginning of the Second Session of the Forty-Third Congress*, vol. 2 (Washington, DC: Government Printing Office, 1875), 507, 511; *Fourteenth Annual Report of the Superintendent of Public Instruction of the State of New York* (Albany: Van Benthuysen and Sons' Steam Printing House, 1868), 33–36, 104–114; *Annual Report, State of New York, 1877*, 115–121.

12. *Annual Report, State of New York, 1868*, 107–109; *Annual Report, State of New York, 1877*, 115–119; Anne Ruggles Gere, "Indian Heart/White Man's Head: Native-American Teachers in Indian Schools, 1880–1930," *History of Education Quarterly* 45, no. 1 (2005): 38–65; Jon Reyner and Jeanne Eder, *American Indian Education: A History* (Norman: University of Oklahoma Press, 2004), 78–80; K. Tsianina Lomawaima and Teresa L. McCarty, *"To Remain an Indian": Lessons in Democracy from a Century of Native American Education* (New York: Teachers College Press, 2006), 77–81. Other sources indicate that local officials were sometimes less than forthcoming about bilingual education to state educational leaders, particularly when it contradicted the state's aims; see Handschin, *The Teaching of Modern Languages*, 67–68.

13. Handschin, *The Teaching of Modern Languages*, 74; Theodore Frelinghuysen Chambers, *The Early Germans of New Jersey: Their History, Churches and Genealogies* (Baltimore: Genealogical Publishing Company, 1969), 34–43; Russell A. Kazal, *Becoming Old Stock: The Paradox of German-American Identity* (Princeton: Princeton University Press, 2004), 1–2; Dieter Cunz, *The Maryland Germans, a History* (Princeton: Princeton University Press, 1948), 11–93.

14. Common Schools of Pennsylvania, *Report of the Superintendent of Common Schools of the Commonwealth of Pennsylvania, for the Year Ending June 3, 1867* (Harrisburg: Singerly and Myers, 1868), xxviii–xliii; Pennsylvania Common Schools, *Report of the Superintendent of Common Schools of Pennsylvania, for the Year Ending June 2, 1857* (Harrisburg: A. Boyd Hamilton, 1858), 9–12; Zeydel, "Teaching of German," 342.

15. *Annual Report, State of Pennsylvania, 1858*, 12–13, 29, 74–75, 132–133; Pennsylvania Common Schools, *Report of the Superintendent of Common Schools of Pennsylvania, for the Year Ending June 3, 1861* (Harrisburg: A. Boyd Hamilton, 1862), 164–165, 200–201, 236–237; Pennsylvania Common Schools, *Report of the Superintendent of Common Schools of Pennsylvania, for the Year Ending June 2, 1862* (Harrisburg: Singerly and Myers, 1863), 182–183, 222, 252–253; Pennsylvania Common Schools, *Report of the Superintendent of Common Schools of Pennsylvania, for the Year Ending June 4, 1863* (Harrisburg: Singerly and Myers, 1864), 196–197, 268–269, 338; Kloss, *The American Bilingual Tradition*, 190; Stempfel, *Fifty Years of Unrelenting German Aspirations in Indianapolis*, 28.

16. *Annual Report, State of Pennsylvania, 1858*, 31; *Annual Report, State of Pennsylvania, 1862*, 43; *Annual Report, State of Pennsylvania, 1863*, 44; *Annual Report, State of Pennsylvania, 1864*, 114.

17. *Forty-Fifth Annual Report of the Board of Commissioners of Public Schools, to the Mayor and City Council of Baltimore, for the Year Ending October 31st, 1873* (Baltimore: King Brothers, 1874), xxxiii; J. G. Morris, "The German in Baltimore," in *Eighth, Ninth and Tenth Annual Reports of the Society for the History of the Germans in Maryland, 1894–1896* (n.p., n.d.), 11–19.

18. Cunz, *The Maryland Germans*, 203–232, 334–338; Morris, "The German in Baltimore," 11–19; Ernest J. Becker, "History of the English-German

Schools in Baltimore," in *Twenty-fifth Report*, Society for the History of the Germans in Maryland (Baltimore: n.p., 1942), 13–17; Douai and Straubenmueller, "German Schools in the United States," 581–586; Carolyn R. Toth, *German-English Bilingual Schools in America: The Cincinnati Tradition in Historical Context* (New York: Peter Lang, 1990), 43–44.

19. Gerber, "Language Maintenance, Ethnic Group Formation, and Public Schools," 31–61; Fessler, "Speaking in Tongues," 35–39; Douai and Straubenmueller, "German Schools in the United States," 581–586; *Forty-Seventh Annual Report of the Board of Commissioners of Public Schools, to the Mayor and City Council of Baltimore, for the Year Ending October 31, 1875* (Baltimore: King Brothers, 1875), xx–xxi, 13, 173; *Forty-Eighth Annual Report of the Board of Commissioners of Public Schools, to the Mayor and City Council of Baltimore, for the Year Ending October 31st, 1876* (Baltimore: King Brothers, 1877), xxxvii–xxxviii, 29; Cunz, *The Maryland Germans*, 231–232, 248–251, 334–338.

20. Newman, "Antebellum School Reform in the Port Cities of the Deep South," 17–35; Kamphoefner et al., *News from the Land of Freedom*, 124–126, 319–322; Carl Bethold, June 3, 1854, in *News from the Land of Freedom*, 329; Hennighausen, "Reminiscences of the Political Life of German-Americans," 4–5.

21. Howard A. M. Henderson, *The Kentucky School-Lawyer: Or A Commentary on the Kentucky School Laws and Regulations of the State Board of Education* (Frankfort, KY: Major, Johnston and Barrett, 1877), 3–11; Barnard, "Subjects and Courses of Instruction in City Public Schools," 537–542; Douai and Straubenmueller, "German Schools in the United States," 591; *Report of the Superintendent of Public Instruction of the Commonwealth of Kentucky, with Summaries of Statistics from the School-Year Ending June 30th, 1881, to the School-Year, 1886* (Frankfort, KY: John D. Woods, Public Printer and Binder, 1886), 54–55; *Report of the Superintendent [of] Public Instruction of the State of Kentucky for Four Scholastic Years Ended June 30, 1891* (Frankfort, KY: E. Polk Johnson, Public Printer and Binder, 1892), 150–152; *Second Report of the Board of Education of Louisville, Kentucky from July 1, 1912 to June 30, 1913* (Louisville: Gross Parsons and Hambleton, n.d.), [ii–iii].

22. Newman, "Antebellum School Reform in the Port Cities of the Deep South," 20–23; Donald E. Devore and Joseph Logsdon, *Crescent City Schools: Public Education in New Orleans, 1841–1991* (The Center for Louisiana Studies, University of Southwestern Louisiana, 1991), 5–25.

23. Newman, "Antebellum School Reform in the Port Cities of the Deep South," 20–23; Devore and Logsdon, *Crescent City Schools*, 25–30.

24. Mark Twain, *Life on the Mississippi* (Mineola, NY: Dover Publications, 2000), 201; Devore and Logsdon, *Crescent City Schools*, 40–55; Louisiana State Department of Education, *Public School Statistics, Session of 1914–'15: Part II of the Biennial Report of 1913–'14 and 1914–'15* (Baton Rouge: Ramires-Jones Printing Company, 1915), 117.

3 Inside the Boardinghouse's Parlor

1. Sally McMurry, "City Parlor, Country Sitting Room: Rural Vernacular Design and the American Parlor, 1840–1900," *Winterhur Portfolio* 20, no. 4 (1985): 261–280; Heinz Kloss, "German-American Language Maintenance Efforts," in *Language Loyalties in the United States: The Maintenance and Perpetuation of Non-English Mother Tongues by American Ethnic and Religious Groups*, ed. Joshua A. Fishman (London: Mouton and Co., 1966), 226.

2. Theodore Dreiser, *A Hoosier Holiday* (1916; reprint, with a new introduction by Douglas Brinkley, Bloomington: Indiana University Press, 1998), 61, 128, 510; Mark Twain, *Life on the Mississippi* (Mineola, NY: Dover Publications, 2000), 254; Henry Boernstein, *The Mysteries of St. Louis*, trans. Friedrich Münch (1851; modern edition by Steven Rowan and Elizabeth Sims, Chicago: Charles H. Kerr Publishing Company, 1990), 111; Theodore Dreiser, *Dawn* (New York: Horace Liveright, Inc., 1931), 7. In the late 1870s, Dreiser's received portions of his elementary education in a German-English Catholic school; see Dreiser, *Dawn*, 27.

3. Dreiser, *A Hoosier Holiday*, 128, 261, 267, 338, 511–512; Twain, *Life on the Mississippi*, 105–106; Kurt Vonnegut, *Bagombo Snuff Box* (New York: Berkley Books, 2000), 357; William J. Reese, "Public Education in Nineteenth-Century St. Louis," in *St. Louis in the Century of Henry Shaw: A View beyond the Garden Wall*, ed. Eric Sandweiss (Columbia: University of Missouri Press, 2003), 167–168.

4. Charles Hart Handschin, *The Teaching of Modern Languages in the United States* (Washington, DC: Government Printing Office, 1913), 67–74; Carolyn R. Toth, *German-English Bilingual Schools in America: The Cincinnati Tradition in Historical Context* (New York: Peter Lang, 1990), 57–58; Paul J. Ramsey, "The War against German-American Culture: The Removal of German-Language Instruction from the Indianapolis Schools, 1917–1919" *Indiana Magazine of History* 98, no. 4 (2002): 290–292; Paul J. Ramsey, "'Let Virtue Be Thy Guide, and Truth Thy Beacon-Light': Moral and Civic Transformation in Indianapolis's Public Schools," in *Civic and Moral Learning in America*, ed. Donald Warren and John J. Patrick (New York: Palgrave Macmillan, 2006), 138–139; Superintendent of Public Instruction, *School Laws of Indiana, as Amended to March 15, 1877, with Opinions, Instructions and Judicial Decisions Relating to Common Schools and the Officers Thereof* (Indianapolis: Sentinel Company, 1877), 48–49.

5. Heinz Kloss, *The American Bilingual Tradition* (1977; reprint, with a new introduction by Reynaldo F. Macías and Terrence G. Wiley, McHenry, IL: Center for Applied Linguistics and Delta Systems Co., 1998), 105–108; Toth, *German-English Bilingual Schools*, 57–58; Ramsey, "The War against German-American Culture," 290–292; Handschin, *The Teaching of Modern Languages*, 67–74; Mary L. Hinsdale, "A Legislative History of the Public School System of the State of Ohio," in *Annual Reports of the Department of Interior for the Fiscal Year Ended June 30, 1901: Report of the Commissioner of Education*, vol. 1 (Washington, DC: Government Printing Office, 1902), 140–155.

6. Handschin, *The Teaching of Modern Languages*, 67–68; Johann Bauer, April 11, 1868, in *News from the Land of Freedom: German Immigrants Write Home*, ed. Walter D. Kamphoefner, Wolfgang Helbich, and Ulrike Sommer, trans. Susan Carter Vogel (Ithaca: Cornell University Press, 1991), 165; Ramsey, "Let Virtue Be Thy Guide," 139.

7. Christian Lenz, January 7, 1868, in *News from the Land of Freedom*, 139; Toth, *German-English Bilingual Schools*, 55–63; Steven L. Schlossman, "Is There an American Tradition of Bilingual Education?: German in the Public Elementary Schools, 1840–1919," *American Journal of Education* 91, no. 2 (1983): 146–147; Common Schools of Cincinnati, *Fifty-Seventh Annual Report for the School Year Ending August 31st, 1886* (Cincinnati: Ohio Valley Publishing and Manufacturing Company, 1887), 93–94; Henry Barnard, ed., "Subjects and Courses of Instruction in City Public Schools," *The American Journal of Education* 3 (1870): 524–532; Hinsdale, "A Legislative History of the Public School System of the State of Ohio," 133–150.

8. Toth, *German-English Bilingual Schools*, 43–44, 55–63; Adolf Douai and John Straubenmueller, "German Schools in the United States," *The American Journal of Education* 3 (1870): 581–586; Ramsey, "Let Virtue Be Thy Guide," 140–143; Schlossman, "Is There an American Tradition of Bilingual Education?," 146–148; William J. Reese, *America's Public Schools: From the Common School to "No Child Left Behind"* (Baltimore: Johns Hopkins University Press, 2005), 86–87.

9. Toth, *German-English Bilingual Schools*, 55–69, 72; Common Schools of Cincinnati, *Thirty-Eighth Annual Report for the School Year Ending June 30, 1867* (Cincinnati: Gazette Steam Book and Job Printing Establishment, 1867), 148–149, 165–195; Common Schools of Cincinnati, *Forty-Second Annual Report for the School Year Ending June 30, 1871* (Cincinnati: Wilstach, Baldwin and Co., 1872), 25–26, 253–257; Common Schools of Cincinnati, *Annual Report, 1887*, 93–96; John B. Peaslee, *Thoughts and Experiences in and out of School* (Cincinnati: Curts and Jennings, 1900), 202–228; John B. Peaslee, "Instruction in German and Its Helpful Influence on Common School Education as Experienced in the Public Schools of Cincinnati" (paper presented at the National German-American Teachers' Association, Chicago, IL, July 19, 1889, University of Illinois Library, Urbana, IL), 8–16; Handschin, *The Teaching of Modern Languages*, 140–142, 147.

10. Frances H. Ellis, "Historical Account of German Instruction in the Public Schools of Indianapolis," *Indiana Magazine of History* 50, no. 2 (1954): 129–130; The Common Schools of Indianapolis, *Annual Report of the Public Schools of the City of Indianapolis, for the School Year Ending Sept. 1, 1866* (Indianapolis: Douglass and Conner, 1867), 7; Ramsey, "Let Virtue Be Thy Guide," 137–140; Paul Rudolph Fessler, "Speaking in Tongues: German-Americans and the Heritage of Bilingual Education in American Public Schools" (PhD diss., Texas A&M University, 1997), 36–39.

11. Ramsey, "Let Virtue Be Thy Guide," 137–143; Kurt Vonnegut, *Palm Sunday: An Autobiographical Collage* (New York: Delta, 1999), 176; *Twenty-Second*

Annual Report of the Public Schools of the City of Indianapolis, for the School Year Ending June 30, 1883 (Indianapolis: Sentinel Co., 1883), 27–31.

12. William J. Reese, " 'Partisans of the Proletariat': The Socialist Working Class and the Milwaukee Schools, 1890–1920," *History of Education Quarterly* 21, no. 1 (1981): 3–50; Bettina Goldberg, "The German-English Academy, the National German-American Teachers' Seminary, and the Public School System in Milwaukee, 1851–1919," in *German Influences on Education in the United States to 1917*, ed. Henry Geitz, Jürgen Heideking, and Jurgen Herbst (Cambridge: Cambridge University Press, 1995), 177–192; Johann Carl Wilhelm Pritzlaff, February 21, 1847, in *News from the Land of Freedom*, 312; Kamphoefner et al., *News from the Land of Freedom*, 299–303; Schlossman, "Is There an American Tradition of Bilingual Education?," 144–145, 164–165, 169–170; Milwaukee Public Schools, *Annual Report of the School Board of the City of Milwaukee, for the Year Ending August 31, 1876* (Milwaukee: n.p., 1876), 8, 57–61, 200–203.

13. Fessler, "Speaking in Tongues," 51–53; Cleveland Public Schools, *Thirty-Seventh Annual Report of the Board of Education for the School Year Ending Aug. 31, 1873* (Cleveland: Leader Printing Company, 1874), 47–65; Cleveland Public Schools, *Thirty-Eighth Annual Report of the Board of Education for the School Year Ending August 31, 1874* (Cleveland: Robison, Savage and Company, 1875), 60–65; L. Viereck, "German Instruction in American Schools," in *Annual Report, Department of Interior, 1902*, vol. 1, 643–644; Kloss, *The American Bilingual Tradition*, 113; "German Schools," *New York Times*, December 8, 1889; L. R. Klemm, *Public Education in Germany and the United States* (Boston: Gorham Press, 1911); Linda L. Sommerfeld, "An Historical Descriptive Study of the Circumstances That Led to the Elimination of German from the Cleveland Schools, 1860–1918" (PhD diss., Kent State University, 1986), 92–100.

14. Jörg Nagler, "Frontier Socialism: The Founding of New Ulm, Minnesota, by German Workers and Freethinkers," in *Emigration and Settlement Patterns of German Communities in North America*, ed. Eberhard Reichmann, LaVern J. Rippley, and Jörg Nagler (Indianapolis: Max Kade German-American Center, 1995), 178–192; Fessler, "Speaking in Tongues," 122–123; Belleville Public Schools, *First Annual Report for the Schoolyear Ending June 26, 1874, and Various Supplementary Documents Exhibiting the Condition of the Schools* (Stern des Westens, 1874), 5, 7–11; *Fifteenth Annual Report of the Belleville Public Schools for the School Year Ending June 22, 1888, and Various Supplementary Documents, Exhibiting the Conditions of the Schools* (Geo. Semmelroth and Co., n.d.), 9–18, 28–29.

15. Wm. Steffen, "German Schools and Teaching German," in *Report of the Secretary of the Interior, Being Part of the Message and Documents Communicated to the Two Houses of Congress at the Beginning of the Third Session of the Forty-First Congress*, vol. 2 (Washington, DC: Government Printing Office, 1870), 437–438; "National German-American Teachers' Association," in *Report of the Secretary of the Interior; Being Part of the Message and Documents Communicated to the Two Houses of Congress*

at the Beginning of the Third Session of the Forty-Sixth Congress, vol. 3 (Washington, DC: Government Printing Office, 1882), 397–398; Viereck, "German Instruction in American Schools," 576–580; C. E. Emmerich, "No Such Sentiment Uttered," *Indianapolis News*, May 3, 1890, 1; *Forty-First Annual Report of the Superintendent of Public Instruction of the State of Michigan, with Accompanying Documents, for the Year 1877* (Lansing: W. S. George and Co., 1878), 139–140; Ramsey, "Let Virtue Be thy Guide," 137–145; Goldberg, "The German-English Academy, the National German-American Teachers' Seminary, and the Public School System in Milwaukee," 177–192; Juliane Jacobi, "Schoolmarm, *Volkserzieher, Kantor,* and *Schulschwester*: German Teachers among Immigrants during the Second Half of the Nineteenth Century," in *German Influences on Education,* 115–128; Toth, *German-English Bilingual Schools,* 67–68.

16. Henry Boernstein, *Memoirs of a Nobody: The Missouri Years of an Austrian Radical, 1849–1866,* trans. and ed. Steven Rowan (St. Louis: Missouri Historical Society, 1997), 221, 228; David A. Gerber, "Language Maintenance, Ethnic Group Formation, and Public Schools: Changing Patterns of German Concern, Buffalo, 1837–1874," *Journal of American Ethnic History* 4, no. 1 (1984): 34–35.

17. Handschin, *The Teaching of Modern Languages,* 67–68; Schlossman, "Is There an American Tradition of Bilingual Education?," 144–146; Toth, *German-English Bilingual Schools,* 76–77; *Thirteenth Biennial Report of the Superintendent of Public Instruction of the State of Illinois: October 1, 1878–June 30, 1880* (Springfield: H. W. Rokker, 1881), 295–296, 303–305.

18. Twain, *Life on the Mississippi,* 274; J. G. Morris, "The German in Baltimore," in *Eighth, Ninth and Tenth Annual Reports of the Society for the History of the Germans in Maryland, 1894–1896* (n.p., n.d.), 12–13; Schlossman, "Is There an American Tradition of Bilingual Education?," 143–144.

19. Kloss, *The American Bilingual Tradition,* 114; Viereck, "German Instruction in American Schools," 644–645; Fessler, "Speaking in Tongues," 38–39; Schlossman, "Is There an American Tradition of Bilingual Education?," 144–145; Barnard, "Subjects and Courses of Instruction in City Public Schools," 524–532, 551–564.

20. Walter D. Kamphoefner, "Learning from the 'Majority-Minority' City: Immigration in Nineteenth-Century St. Louis," in *St. Louis in the Century of Henry Shaw,* 79–99; Reese, "Public Education in Nineteenth-Century St. Louis," 167–187; Henry Barnard, "William Torrey Harris and St. Louis Public Schools," *American Journal of Education* 30 (1880): 635; *Twenty-First Annual Report of the Board of Directors of the St. Louis Public Schools, for the Year Ending August 1, 1875* (St. Louis: Globe-Democrat Job Printing Co., 1876), 114–116; *Twenty-Fifth Annual Report of the Board of Directors of the St. Louis Public Schools, for the Year Ending August 1, 1879* (St. Louis: G. I. Jones and Co., 1880), 98–99; Selwyn K. Troen, *The Public and the Schools: Shaping the St. Louis System, 1830–1920* (Columbia: University of Missouri Press, 1975), 55–78.

21. *Seventeenth Annual Report of the Board of Directors of the St. Louis Public Schools, for the Year Ending August 1, 1871* (St. Louis: Plate, Olshausen

and Co., 1872), 103; Kloss, "German-American Language Maintenance Efforts," 226; W. T. Harris, "German Instruction in American Schools and the National Idiosyncrasies of the Anglo-Saxons and the Germans" (paper presented at the National German-American Teachers' Association, Cleveland, OH, July 16, 1890, University of Illinois Library, Urbana, IL), 1–2. Italics added.

22. *Annual Report, St. Louis, 1876*, 114–121; Troen, *The Public and the Schools*, 64; Kamphoefner, "Learning from the 'Majority-Minority' City," 97; Harris, "German Instruction in American Schools," 2.

23. Reese, "Public Education in Nineteenth-Century St. Louis," 169–177; William J. Reese, "The Philosopher-King of St. Louis," in *Curriculum and Consequence: Herbert M. Kliebard and the Promise of Schooling*, ed. Barry M. Franklin (New York: Teachers College Press, 2000), 155–177; *Annual Report, St. Louis, 1872*, 103, 109; *Annual Report, St. Louis, 1876*, 114–127.

24. Harris, "German Instruction in American Schools," 1–2; Barnard, "William Torrey Harris and St. Louis Public Schools," 635; Reese, "The Philosopher-King of St. Louis," 155–177; Selma Cantor Berrol, *Growing Up American: Immigrant Children in America, Then and Now* (New York: Twayne Publishers, 1995), 91–95.

25. *Thirty-Ninth Annual Report of the Superintendent of Public Instruction of the State of Michigan, with Accompanying Documents, for the Year 1875* (Lansing: W. S. George and Co., 1876), 58–59.

26. Department of the Interior, Census Office, *Statistics of the Population of the United States at the Tenth Census (June 1, 1880)* (Washington, DC: Government Printing Office, 1883), 470; Olaf Morgan Norlie, *History of the Norwegian People in America* (Minneapolis: Augsburg Publishing House, 1925), 234–236, 312–313; Odd S. Lovell, *The Promise of America: A History of the Norwegian-American People* (Minneapolis: University of Minnesota Press, 1984), 28, 34.

27. O. E. Rolvaag, *Giants in the Earth* (New York: Harper and Brothers Publishers, 1929), 258; Einar Haugen, *The Norwegian Language in America: A Study in Bilingual Behavior*, vol. 1 (Philadelphia: University of Pennsylvania Press, 1953), 101–102; Norlie, *History of the Norwegian People*, 214–216; Andreas Hjerpeland to Ivar Kleiven, May 3, 1877, in *In Their Own Words: Letters from Norwegian Immigrants*, trans. and ed. Solveig Zempel (Minneapolis: University of Minnesota Press, 1991), 11–12; Andreas Hjerpeland to Ivar Kleiven, December 7, 1878, in *In Their Own Words*, 16.

28. Haugen, *The Norwegian Language in America*, vol. 1, 101–110.

29. Rolvaag, *Giants in the Earth*, 220–221; Andreas Hjerpeland to Ivar Kleiven, June 24, 1871, in *In Their Own Words*, 3–4; Andreas Hjerpeland to Ivar Kleiven, July 7, 1876, in *In Their Own Words*, 10–11; Andreas Hjerpeland to Ivar Kleiven, May 3, 1877, in *In Their Own Words*, 11–12; Andreas Hjerpeland to Ivar Kleiven, July 24, 1878, in *In Their Own Words*, 14–15; Andreas Hjerpeland to Ivar Kleiven, February 24, 1893, in *In Their Own Words*, 22–23; J. J. Skordalsvold, "Historical Review of the Scandinavian

Schools in Minnesota," in *History of the Scandinavians and Successful Scandinavians in the United States*, ed. O. N. Nelson, vol. 1, 2nd ed. (Minneapolis: O. N. Nelson and Company, 1900), 317–334; Andrew Estrem, "Historical Review of Luther College," in *History of the Scandinavians and Successful Scandinavians*, vol. 2, 23–37; J. J. Skordalsvold, "Historical Review of the Scandinavian Schools in Wisconsin," in *History of the Scandinavians and Successful Scandinavians*, vol. 2, 129–134; Haugen, *The Norwegian Language in America*, vol. 1, 100–103.

4   The Polyglot Boarders Move West

1. Theodore Dreiser, *A Hoosier Holiday* (1916; reprint, with a new introduction by Douglas Brinkley, Bloomington: Indiana University Press, 1998), 61; Manuel G. Gonzales, *Mexicanos: A History of Mexicans in the United States* (Bloomington: Indiana University Press, 2000), 72–81; James K. Polk, "Expanding Westward," in *Our Nation's Archive: The History of the United States in Documents*, ed. Erik Bruun and Jay Crosby (New York: Tess Press, 1999), 314.

2. Polk, "Expanding Westward," in *Our Nation's Archive*, 314; Johann Carl Wilhelm Pritzlaff, February 21, 1847, in *News from the Land of Freedom: German Immigrants Write Home*, ed. Walter D. Kamphoefner, Wolfgang Helbich, and Ulrike Sommer, trans. Susan Carter Vogel (Ithaca: Cornell University Press, 1991, 313; Gonzales, *Mexicanos*, 75–81, 91–92; *Atlas of American History* (Skokie, IL: Houghton Mifflin / Rand McNally, 1993), 27–29.

3. Francis Parkman, *The Oregon Trail*, ed. Mason Wade (Heritage Press, 1943), 4, 33, 254; Elliot West, *Growing Up with the Country: Childhood on the Far Western Frontier* (Albuquerque: University of New Mexico Press, 1989), 2–7; Gonzales, *Mexicanos*, 106–108.

4. West, *Growing Up with the Country*, 2–7; Gonzales, *Mexicanos*, 62–69, 84–91; Parkman, *The Oregon Trail*, 1, 189–196; Richard Henry Dana, Jr., *Two Years before the Mast*, ed. Thomas Philbrick (New York: Penguin Books, 1981), 124–125, 237; Carl Blümner, March 18, 1841, in *News from the Land of Freedom*, 105; Cormac McCarthy, *Blood Meridian, or the Evening Redness in the West* (New York: Modern Library, 2001), 78.

5. West, *Growing Up with the Country*, 12–21, 190.

6. Frederick C. Luebke, "Ethnic Group Settlement on the Great Plains," *The Western Historical Quarterly* 8, no. 4 (1977): 405–430; Parkman, *The Oregon Trail*, 9, 63, 92, 97–113; Mark Twain, *Life on the Mississippi* (Mineola, NY: Dover Publications, 2000), 270.

7. Luebke, "Ethnic Group Settlement on the Great Plains," 405–430; Andreas Hjerpeland to Ivar Kleiven, June 24, 1871, in *In Their Own Words: Letters from Norwegian Immigrants*, ed. and trans. Solveig Zempel (Minneapolis: University of Minnesota Press, 1991), 4; Andreas Hjerpeland to Ivar Kleiven, February 24, 1893, in *In Their Own Words*, 22–23; Einar Haugen, *The Norwegian Language in America: A Study of Bilingual Behavior*, vol. 1

(Philadelphia: University of Pennsylvania Press, 1953), 18–27; Olaf Morgan Norlie, *History of the Norwegian People in America* (Minneapolis: Augsburg Publishing House, 1925), 82–83; Walter Struve, *Germans and Texans: Commerce, Migration, and Culture in the Days of the Lone Star Republic* (Austin: University of Texas Press, 1996), 39–66; Carlos Kevin Blanton, *The Strange Career of Bilingual Education in Texas, 1836–1981* (College Station: Texas A&M University Press, 2004), 31–37; O. E. Rolvaag, *Giants in the Earth* (New York: Harper and Brothers Publishers, 1929), 129.

8. Luebke, "Ethnic Group Settlement on the Great Plains," 405–430; West, *Growing Up with the Country*, 179–210; Rolvaag, *Giants in the Earth*, 257–258; *Eighth Biennial Report of the Superintendent of Public Instruction: State of South Dakota, 1905–1906* (Huron, SD: John Longstaff, n.d.), 5–7, 64. Soddies were common farther south as well. In 1897, for example, Nebraska still had 622 sod schoolhouses in use; see *Fifteenth Biennial Report of the State Superintendent of Public Instruction to the Governor of the State of Nebraska, Dec. 1, 1898* (Lincoln: Jacob North and Co., 1898), 360–361.

9. Charles Hart Handschin, *The Teaching of Modern Languages in the United States* (Washington, DC: Government Printing Office, 1913), 67; [*Third Biennial Report of the Superintendent of Public Instruction to the Governor of North Dakota, 1892–1894*] (n.p., n.d.), 137–138, 149–150; *Fourth Biennial Report of the Superintendent of Public Instruction to the Governor of North Dakota, for the Two Years Ending June 30, 1896* (Jamestown, ND: Alert, State Printers and Binders, 1896), 16, 283–284.

10. Stephen Crane, *The Blue Hotel*, ed. Joseph Katz (Columbus: Charles E. Merrill Publishing Company, 1969), 13; Luebke, "Ethnic Group Settlement on the Great Plains," 405–430; Heinz Kloss, *The American Bilingual Tradition* (1977; reprint, with a new introduction by Reynaldo F. Macias and Terrence G. Wiley, McHenry, IL: Center for Applied Linguistics and Delta Systems Co., 1998), 107; Blanton, *Strange Career*, 18–21, 31–37; *School Laws of Texas, 1913* (Austin: Von [Boe]ckmann-Jones Company, n.d.), 82–83; Lauren Ann Kattner, "Land and Marriage: German Regional Reflections in Four Texas Towns, 1845–1860," in *Emigration and Settlement Patterns of German Communities in North America*, ed. Eberhard Reichmann, LaVern J. Rippley, and Jörg Nagler (Indianapolis: Max Kade German-American Center, 1995), 237–257; Struve, *Germans and Texans*, 39–66.

11. Gonzalez, *Mexicanos*, 58–69, 96; Gabriel García Márquez, *One Hundred Years of Solitude*, trans. Gregory Rabassa (New York: Perennial Classics, 1998), 181, 361; George I. Sanchez, *Forgotten People: A Study of New Mexicans* (Albuquerque: University of New Mexico Press, 1940), 10–12.

12. Gonzales, *Mexicanos*, 64–69; Guadalupe Vallejo, "The Romantic Frontier," in *Foreigners in Their Native Land: Historical Roots of the Mexican Americans*, ed. David J. Weber, 30th anniversary ed. (Albuquerque: University of New Mexico Press, 2003), 45–49; Willa Cather, *Death Comes for the Archbishop* (1925; reprint, New York: Vintage Classics, 1990), 153.

13. Dana, *Two Years before the Mast*, 126–127; Frederick Douglass, *Narrative of the Life of Frederick Douglass: An American Slave* (New York: Signet

Classic, 1997), 19–21; Ronald Takaki, *A Different Mirror: A History of Multicultural America* (Boston: Little, Brown and Company, 1993), 121–126.

14. Stephen Watts Kearney, "Their Property, Their Persons, Their Religion," in *Foreigners in Their Native Land*, 161–162; Treaty of Guadalupe Hidalgo, Article VIII, in *Foreigners in Their Native Land*, 163–164; John C. Calhoun, "The Government of a White Race," in *Foreigners in Their Native Land*, 135–137; Walter Prescott Webb, "Blood…as Ditch Water," in *Foreigners in Their Native Land*, 76–77; Parkman, *Oregon Trail*, 240.

15. Gonzales, *Mexicanos*, 82–112; Stuart Ewen and Elizabeth Ewen, *Typecasting: On the Arts and Sciences of Human Inequality* (New York: Seven Stories Press, 2006), 11–17, 39; Dana, *Two Years before the Mast*, 123, 237; Rufus B. Sage, "Degenerate Inhabitants of New Mexico," in *Foreigners in Their Native Land*, 71–75; Parkman, *Oregon Trail*, 238; Orhan Pamuk, *Snow*, trans. Maureen Freely (New York: Vintage International, 2005), 275–276; José María Sánchez, "A Trip to Texas in 1828," in *Foreigners in Their Native Land*, 78–83.

16. John M. Nieto-Phillips, *The Language of Blood: The Making of Spanish-American Identity in New Mexico, 1880s-1930s* (Albuquerque: University of New Mexico Press, 2004), 1–11, 171–197; Sanchez, *Forgotten People*, 15–26; Vallejo, "The Romantic Frontier," in *Foreigners in Their Native Land*, 46; Gonzales, *Mexicanos*, 111–112.

17. Nieto-Phillips, *The Language of Blood*, 5–11; Sanchez, *Forgotten People*, 10–14; Sánchez, "A Trip to Texas in 1828," in *Foreigners in Their Native Land*, 78–81; Gonzales, *Mexicanos*, 84–125; Annie Reynolds, *The Education of Spanish-Speaking Children in Five Southwestern States* (Washington, DC: Government Printing Office, 1933), 5–8; W. W. Mills, *Forty Years at El Paso, 1858–1898* (El Paso: Carl Hertzog, 1962), 9.

18. Gonzales, *Mexicanos*, 70; Carl Blümner, March 18, 1841, in *News from the Land of Freedom*, 105; Dana, *Two Years before the Mast*, 128, 130; Sánchez, "A Trip to Texas in 1828," in *Foreigners in Their Native Land*, 81.

19. Treaty of Guadalupe Hidalgo, Article IX, in *Foreigners in Their Native Land*, 164; Kloss, *The American Bilingual Tradition*, 160–163; Blanton, *Strange Career*, 26–28; Nieto-Phillips, *The Language of Blood*, 72–80; Robert R. Calderón, "Unión, Paz y Trabajo: Laredo's Mexican Mutual Aid Societies, 1890s," in *Mexican Americans in Texas History: Selected Essays*, ed., Emilio Zamora, Cynthia Orozco, and Rodolfo Rocha (Austin: Texas State Historical Association, 2000), 66–70; *Sixth Biennial Report of the Superintendent of Public Instruction for the Scholastic Years Ending August 31, 1887, and July 1, 1888, Being the Thirteenth Report from the Department of Education* (Austin: State Printing Office, 1888), 301.

20. Kloss, *The American Bilingual Tradition*, 229–240; West, *Growing Up with the Country*, 190; *Biennial Report of the Superintendent of Public Instruction of the Territory of Arizona* (n.p., 1890), 32–36; *Sixth Biennial Report of the Superintendent of Public Instruction of the State of Colorado, for Biennial Term Ending June 30, 1888* (Denver: Collier and Cleaveland,

1889), 67–69; California Council of Education, *First Report of the Committee on the Course of Study for Elementary Schools* (Oakland: n.p., 1896), 27.

21. Blanton, *Strange Career*, 18–31; Mills, *Forty Years at El Paso*, 5, 23.

22. Blanton, *Strange Career*, 28–30; *Biennial Report, Texas, 1888*, 297, 301, 309.

23. Gonzales, *Mexicanos*, 98–106; Nieto-Phillips, *The Language of Blood*, 1–11, 171–205; "The Spanish Language and Its Importance for the Natives of New Mexico," in *Foreigners in Their Native Land*, 254–255.

24. West, *Growing Up with the Country*, 190; Kloss, *The American Bilingual Tradition*, 169–171; Nieto-Phillips, *The Language of Blood*, 80, 197–205; *Seventeenth and Eighteenth Annual Reports of the Territorial Superintendent of Public Instruction to the Governor of New Mexico for the Years 1907–1908* (Santa Fe: New Mexican Printing Company, 1909), 11, 17–18.

25. Alexis de Tocqueville, *Democracy in America and Two Essays on America*, ed. Isaac Kramnick, trans. Gerald E. Bevan (New York: Penguin Books, 2003), 324, 330–331, 380; "President Jackson on Indian Removal: December 8, 1829," in *Documents of United States Indian Policy*, ed. Francis Paul Prucha, 3rd ed. (Lincoln: University of Nebraska Press, 2000), 47–48; "Indian Removal Act: May 28, 1830," in *Documents of United States Indian Policy*, 52–53; Takaki, *A Different Mirror*, 88–98; Jon Reyhner and Jeanne Eder, *American Indian Education: A History* (Norman: University of Oklahoma Press, 2004), 48–51; Francis A. Walker, *The Statistics of the Population of the United States, Embracing the Tables of Race, Nationality, Sex, Selected Ages, and Occupations*, ninth census, vol. 1 (Washington, DC: Government Printing Office, 1872), xvii; *Report of the Secretary of the Interior, Being Part of the Message and Documents Communicated to the Two Houses of Congress at the Beginning of the Third Session of the Forty-First Congress*, vol. 2 (Washington, DC: Government Printing Office, 1870), 348–352.

26. John McWhorter, *The Power of Babel: A Natural History of Language* (New York: Perennial, 2003), 258; Takaki, *A Different Mirror*, 98–100, 238–240; Reyhner and Eder, *American Indian Education*, 51–58, 130–131; Francis La Flesche, *The Middle Five: Indian Schoolboys of the Omaha Tribe* (1900; reprint, with an introduction by David A. Baerreis, Madison: University of Wisconsin Press, 1963), viii, 15; Parkman, *The Oregon Trail*, 153, 189; Forrest Carter, *The Education of Little Tree* (Albuquerque: University of New Mexico Press, 1976), 145; K. Tsianina Lomawaima and Teresa L. McCarty, *"To Remain an Indian": Lessons in Democracy from a Century of Native American Education* (New York: Teachers College Press, 2006), 91–115; James Wilson, *The Earth Shall Weep: A History of Native America* (New York: Grove Press, 1998), 332–326.

27. Parkman, *The Oregon Trail*, 153, 177, 182, 184, 189, 224; *Report of the Secretary of the Interior; Being Part of the Message and Documents Communicated to the Two Houses of Congress at the Beginning of the Second Session of the Forty-Sixth Congress*, vol. 1 (Washington, DC: Government

Printing Office, 1879), 4; Rogers M. Smith, *Civic Ideals: Conflicting Visions of Citizenship in U.S. History* (New Haven: Yale University Press, 1997), 318–320; Herman Melville, *The Confidence Man* (New York: Prometheus Books, 1995), 176–185; Cather, *Death Comes for the Archbishop*, 183–184, 290–292.

28. David Wallace Adams, *Education for Extinction: American Indians and the Boarding School Experience, 1875–1928* (Lawrence: University Press of Kansas, 1995), 6–9; Reyhner and Eder, *American Indian Education*, 59–80; Parkman, *The Oregon Trail*, 94; "Secretary of the Interior Cox on Reservations and on the Peace Policy," in *Documents of United States Indian Policy*, 128–130; "Establishment of a Reservation by Executive Order: May 29, 1873," in *Documents of United States Indian Policy*, 142; *Report of the Secretary of the Interior, 1870*, vol. 2, 339–340; *Report of the Secretary of the Interior, 1879*, vol. 1, "Map Showing Indian Reservations."

29. "Establishment of the War Department: August 7, 1789," in *Documents of United States Indian Policy*, 13–14; "Transfer of Indian Affairs to the Department of the Interior: March 3, 1849," in *Documents of United States Indian Policy*, 79–80; *Report of the Secretary of the Interior, 1870*, vol. 2, 340; Wilson, *The Earth Shall Weep*, 289–297; Smith, *Civic Ideals*, 318–320; Adams, *Education for Extinction*, 7–18; *Report of the Secretary of the Interior, 1879*, vol. 1, 4.

30. Adams, *Education for Extinction*, 7–27; David Wallace Adams, "Land, Law, and Education: The Troubled History of Indian Citizenship, 1871–1924," in *Civic and Moral Learning in America*, ed. Donald Warren and John J. Patrick (New York: Palgrave Macmillan, 2006), 119–134; *Report of the Secretary of the Interior, 1879*, vol. 1, 5; G. Stanley Hall, "The White Man's Burden Versus Indigenous Development for the Lower Races," in *Journal of Proceedings and Addresses of the National Educational Association* (Chicago: University of Chicago Press, 1903), 1054.

31. Adams, *Education for Extinction*, 28–59; Reyhner and Eder, *American Indian Education*, 51–80; Francis Paul Prucha, *The Churches and the Indian Schools, 1888–1912* (Lincoln: University of Nebraska Press, 1979), 1–40; *Report of the Secretary of the Interior, 1879*, vol. 1, 10–11; *Report of the Secretary of the Interior; Being Part of the Message and Documents Communicated to the Two Houses of Congress at the Beginning of the Second Session of the Forty-Eighth Congress*, vol. 2 (Washington, DC: Government Printing Office, 1884), 501–502, 787–788; "Assignment of Indian Agencies to Religious Societies: November 1, 1872," in *Documents of United States Indian Policy*, 140–142; Hall, "The White Man's Burden," 1053–1056.

32. *Report of the Secretary of the Interior, 1879*, vol. 1, 10; Reyhner and Eder, *American Indian Education*, 74–78; *Report of the Secretary of the Interior; Being Part of the Message and Documents Communicated to the Two Houses of Congress at the Beginning of the First Session of the Fiftieth Congress*, vol. 2 (Washington, DC: Government Printing Office, 1887), 761, 783; "Supplemental Report on Indian Education: December 1, 1889," in *Documents of United States Indian Policy*, 176–179.

33. *Report of the Secretary of the Interior, 1884,* vol. 2, 787; La Flesche, *The Middle Five,* xvii; Luther Standing Bear, *My People the Sioux,* ed. E. A. Brininstool (1928; reprint, with an introduction by Richard N. Ellis, Lincoln: University of Nebraska Press, 1975), 136–137.

34. Arnold H. Leibowitz, "A History of Language Policy in American Indian Schools," in *Bilingual Education for American Indians* (New York: Arno Press, 1978), 1–4; Reyhner and Eder, *American Indian Education,* 44–80; Parkman, *The Oregon Trail,* 252; *Report of the Secretary of the Interior, 1887,* vol. 2, 763; Prucha, *The Churches and the Indian Schools,* 1–40.

35. Adams, *Education for Extinction,* 48–59; Richard Henry Pratt, *Battlefield and Classroom: Four Decades with the American Indian, 1867–1904,* ed. Robert M. Utley (New Haven: Yale University Press, 1964), 238; *Report of the Secretary of the Interior, 1879,* vol. 1, 73–74; Standing Bear, *My People the Sioux,* 136–137, 192–193; Reyhner and Eder, *American Indian Education,* 52–58; La Flesche, *The Middle Five,* 3, 13, 22.

36. Ronald Takaki, *Strangers from a Different Shore: A History of Asian Americans* (Boston: Little, Brown and Company, 1989), 99–112; *Report of the Secretary of the Interior, 1870,* vol. 2, 427–428; House Committee on Education and Labor, *Chinese Immigration,* 45th Cong., 2d sess., 1878, H. Rep. 240, 1–5.

37. Committee of the Senate of California, *An Address to the People of the United States upon the Social, Moral, and Political Effect of Chinese Immigration,* 45th Cong., 1st sess., 1877, H. Mis. Doc. 9, 22; Smith, *Civic Ideals,* 358–365; Takaki, *Strangers from a Different Shore,* 110–112; Takaki, *A Different Mirror,* 204–209.

38. Takaki, *Strangers from a Different Shore,* 79–131; Takaki, *A Different Mirror,* 191–221; *Report of the Secretary of the Interior, 1870,* vol. 2, 422–423; House Committee on Education and Labor, *Chinese Immigration,* 1–5.

39. Takaki, *Strangers from a Different Shore,* 105–112; Takaki, *A Different Mirror,* 191–221; Victor Low, *The Unimpressible Race: A Century of Educational Struggle by the Chinese in San Francisco* (San Francisco: East/West Publishing Company, Inc., 1982), 6–37; Committee on Education and Labor, *Chinese Immigration,* 2; Ralph Ellison, *Invisible Man* (New York: Vintage International, 1995), 3, 507; Committee of the Senate of California, *An Address to the People of the United States,* 22, 34.

40. Low, *The Unimpressible Race,* 13–37, 54–84; *Report of the Secretary of the Interior, 1870,* vol. 2, 430–432; Takaki, *Strangers from a Different Shore,* 102–103; John Swett, *Public Education in California* (1911; reprint, New York: Arno Press, 1969), 106–107, 185–187.

41. Low, *The Unimpressible Race,* 13–37, 56, 77–84, 221; John Cotter Pelton, *Life's Sunbeams and Shadows: Poems and Prose with Appendix, Including Biographical and Historical Notes in Prose,* vol. 1 (San Francisco: Bancroft Company, 1893), 213–224; Swett, *Public Education in California,* 106–107; Takaki, *Strangers from a Different Shore,* 120–121, 127–129.

5   Nativism among the Homeowners: The Metaphysics of Foreigner-Hating

1. Herman Melville, *The Confidence Man* (New York: Prometheus Books, 1995), 176–189; Francis Parkman, *The Oregon Trail*, ed. Mason Wade (Heritage Press, 1943), 224, 284; Ronald Takaki, *Iron Cages: Race and Culture in 19th-Century America*, revised ed. (New York: Oxford University Press, 2000), 3, 80–107; Andrew Jackson, "First Annual Message," in *Documents of American Prejudice: An Anthology of Writings on Race from Thomas Jefferson to David Duke*, ed. S. T. Joshi (New York: Basic Books, 1999), 238; Charles Peirce, "The Fixation of Belief," in *The American Intellectual Tradition: A Sourcebook*, ed. David A. Hollinger and Charles Capper, vol. 2, 4th ed. (New York: Oxford University Press, 2001), 14. Takaki also draws upon Melville's phraseology to explore the nature of American racism.
2. Ronald Takaki, *A Different Mirror: A History of Multicultural America* (Boston: Little, Brown and Company, 1993), 79–80, 177–178; Juan Nepomuceno Seguín, "A Foreigner in My Native Land," in *Foreigners in Their Native Land: Historical Roots of the Mexican Americans*, ed. David J. Weber, 30th anniversary ed. (Albuquerque: University of New Mexico Press, 2003), 178; Louis P. Hennighausen, "Reminiscences of the Political Life of the German-Americans in Baltimore during the Years 1850–1860," in *Eleventh and Twelfth Annual Reports of the Society for the History of the Germans in Maryland, 1897–1898* (n.p., n.d.), 7, 16–17.
3. Rogers M. Smith, *Civic Ideals: Conflicting Visions of Citizenship in U.S. History* (New Haven: Yale University Press, 1997), 5–39; Benjamin Rush, "Thoughts upon the Mode of Education Proper in a Republic," in *The School in the United States: A Documentary History*, ed. James W. Fraser (Boston: McGraw-Hill, 2001), 28; Marcius Willson, *The Fourth Reader of the United States Series* (New York: Harper and Brothers, 1872), iii, 22; Benjamin B. Comegys, ed., *A Primer of Ethics* (Boston: Ginn and Company, 1891), 113, 124; Ruth Miller Elson, *Guardians of Tradition: American Schoolbooks of the Nineteenth Century* (Lincoln: University of Nebraska Press, 1964), 41–43; H. L. Mencken, "Puritanism as a Literary Force," in *The American Intellectual Tradition*, 196; Julie A. Reuben, *The Making of the Modern University: Intellectual Transformation and the Marginalization of Morality* (Chicago: University of Chicago Press, 1996), 133–175; Paul J. Ramsey, " 'Let Virtue Be Thy Guide, and Truth Thy Beacon-Light': Moral and Civic Transformation in Indianapolis's Public Schools," in *Civic and Moral Learning in America*, ed. Donald Warren and John J. Patrick (New York: Palgrave, 2006), 135–151.
4. John Higham, *Strangers in the Land: Patterns of American Nativism, 1860–1925* (New Brunswick: Rutgers University Press, 1955), 4–18.
5. Frederick Jackson Turner, "The Significance of the Frontier in American History," in *The American Intellectual Tradition*, 84–92; John O'Sullivan, "The Flight from Famine in Ireland," in *Immigration as a Factor in American History*, ed. Oscar Handlin (Englewood Cliffs, NJ: Prentice-Hall Inc., 1959), 21–22; Lord Mountcashell, "The Flight from Famine in Ireland,"

in *Immigration as a Factor in American History*, 22–24; "Migration out of Germany," in *Immigration as a Factor in American History*, 24–26; Robert H. Wiebe, *The Segmented Society: An Introduction to the Meaning of America* (New York: Oxford University Press, 1975), 20; B. Edward McClellan, *Moral Education in America: Schools and the Shaping of Character from Colonial Times to the Present* (New York: Teachers College Press, 1999), 15–45.

6. McClellan, *Moral Education in America*, 15–45; David Riesman, in collaboration with Reuel Denney and Nathan Glazer, *The Lonely Crowd: A Study of the Changing American Character* (New Haven: Yale University Press, 1950), 11–16; Willson, *The Fourth Reader of the United States*, 22, 29, 53–54, 162–163; Elson, *Guardians of Tradition*, 41; Mark Twain, *Life on the Mississippi* (Mineola, NY: Dover Publications, Inc., 2000), 246; Mencken, "Puritanism as a Literary Force," in *The American Intellectual Tradition*, 195–198.

7. Philip Greven, *The Protestant Temperament: Patterns of Child-Rearing, Religious Experience, and the Self in Early America* (Chicago: University of Chicago Press, 1988), 110; Steven Marcus, *The Other Victorians: A Study of Sexuality and Pornography in Mid-Nineteenth-Century England* (New York: Basic Books, Inc., 1966), 283–284; Erich Fromm, *Escape from Freedom* (New York: Henry Holt and Company, 1994), 46; Alexis de Tocqueville, *Democracy in America and Two Essays on America*, ed. Isaac Kramnick, trans. Gerald E. Bevan (New York: Penguin Books, 2003), 294–297; Lawrence J. Friedman, *Inventors of the Promised Land* (New York: Alfred A. Knopf, 1975), 307–314; Peirce, "The Fixation of Belief," in *The American Intellectual Tradition*, 11.

8. "Letter from a Rabid Know-Nothing—Views, Feelings, and Aims of the Party," *New York Daily Times*, June 8, 1854, 2; Hennighausen, "Reminiscences of the Political Life of the German-Americans," 16; Washington Gladden, "The Anti-Catholic Crusade," *Century* 47 (March 1894): 789–795; Higham, *Strangers in the Land*, 5–7; John J. Appel and Selma Appel, "The Huddled Masses and the Little Red Schoolhouse," in *American Education and the European Immigrant, 1840–1940*, ed. Bernard J. Weiss (Urbana: University of Illinois Press, 1982), 23–24.

9. John T. McGreevy, *Catholicism and American Freedom: A History* (New York: W. W. Norton and Company, 2003), 19–26; Higham, *Strangers in the Land*, 4–7, 31–32; "Letter from a Rabid Know-Nothing," 2; "Petition of the Catholics of New York for a Portion of the Common School Fund: To the Honorable Board of Aldermen of the City of New York, 1840," in *The School in the United States*, 74–80.

10. Henry Boernstein, *Memoirs of a Nobody: The Missouri Years of an Austrian Radical, 1849–1866*, ed. and trans. Steven Rowan (St. Louis: Missouri Historical Society Press, 1997), 177, 204; Higham, *Strangers in the Land*, 5–7; Wiebe, *The Segmented Society*, 32, 65–67; George M. Stephenson, "Nativism in the Forties and Fifties, with Special Reference to the Mississippi Valley," *The Mississippi Valley Historical Review* 9, no. 3 (1922): 185–202; Carl Berthold, February 23, 1853, in *News from the Land*

of Freedom: German Immigrants Write Home, ed. Walter D. Kamphoefner, Wolfgang Helbich, and Ulrike Sommer, trans. Susan Carter Vogel (Ithaca: Cornell University Press, 1991), 325; "America Belongs to Americans," in *Immigration: Opposing Viewpoints*, ed. Teresa O'Neill (San Diego: Greenhaven Press, 1992), 73–76; "Letter from a Rabid Know-Nothing," 2.

11. Boernstein, *Memoirs of a Nobody*, 178–180; Walter D. Kamphoefner, Wolfgang Helbich, and Ulrike Sommer, eds., *News from the Land of Freedom: German Immigrants Write Home*, trans. Susan Carter Vogel (Ithaca: Cornell University Press, 1991), 126; Christian Lenz, May 29, 1855, in *News from the Land of Freedom*, 133; Hennighausen, "Reminiscences of the Political Life of the German-Americans," 9–11.

12. Boernstein, *Memoirs of a Nobody*, 207–208; Elson, *Guardians of Tradition*, 47; McGreevy, *Catholicism and American Freedom*, 7–11, 22–23, 38–42; John Swett, *Public Education in California* (1911; reprint, New York: Arno Press, 1969), 115–117; Higham, *Strangers in the Land*, 5–7. Although not nativists, many of the freethinking leaders of German-English schooling were anti-Catholics themselves because of their negative experiences with the Church in Europe; see Henry Boernstein, *The Mysteries of St. Louis*, trans. Friedrich Münch (1851; reprint, edited by Steven Rowan and Elizabeth Sims, Chicago: Charles H. Kerr Publishing Company, 1990), 63–67, 92–94; Clemens Vonnegut, *A Proposed Guide for Instruction in Morals from the Standpoint of a Freethinker for Adult Persons Offered by a Dilettante* (1900; reprint, n.p., 1987), 5–10.

13. *Eleventh Biennial Report of the Superintendent of Public Instruction of the State of Illinois for the Two Years Ending Sept. 30, 1876* (Springfield: D. W. Lusk, State Printers and Binders, 1877), 341; Ludwig Dilger, March 23, [1913?], in *News from the Land of Freedom*, 502; Higham, *Strangers in the Land*, 28–30, 61–65; Donald L. Kinzer, *An Episode of Anti-Catholicism: The American Protective Association* (Seattle: University of Washington Press, 1964), 3–57; Gladden, "The Anti-Catholic Crusade," 789–795.

14. *Minority Report of the Committee on Text-Books, on Text-Books in History in the High Schools* (Boston: Rockwell and Churchill, City Printers, 1890), 3–6; Franz Joseph Löwen, [1883], in *News from the Land of Freedom*, 198; "Third Plenary Council of Baltimore, 1884," in *The School in the United States*, 145; Lloyd P. Jorgenson, *The State and the Non-Public School, 1825–1925* (Columbia: University of Missouri Press, 1987), 70–136; James W. Fraser, *Between Church and State: Religion and Public Education in a Multicultural America* (New York: St. Martin's Griffin, 2000), 57–65; Appel and Appel, "The Huddled Masses and the Little Red Schoolhouse," 23–24.

15. N. C. Dougherty, "Recent Legislation upon Compulsory Education in Illinois and Wisconsin," in *Journal of Proceeding and Addresses* (New York: National Educational Association, 1891), 393–401; Jorgenson, *The State and the Non-Public School*, 188–204; Victor Low, *The Unimpressible Race: A Century of Educational Struggle by the Chinese in San Francisco* (San Francisco: East/West Publishing Company, 1982), 59–91; Jon Reyhner and Jeanne Eder, *American Indian Education: A History* (Norman: University

of Oklahoma Press, 2004), 74–80; Francis Paul Prucha, *The Churches and the Indian Schools, 1888–1912* (Lincoln: University of Nebraska Press, 1979), 1–40; Gladden, "The Anti-Catholic Crusade," 790.

16. Turner, "The Significance of the Frontier," in *The American Intellectual Tradition*, 92; Tocqueville, *Democracy in America*, 728; Melville, *The Confidence Man*, 67–72; Henry Steele Commager, *The American Mind: An Interpretation of American Thought and Character Since the 1880's* (New Haven: Yale University Press, 1950), 5; Friedman, *Inventors of the Promised Land*, 310.

17. Stuart Ewen and Elizabeth Ewen, *Typecasting: On the Arts and Sciences of Human Inequality* (New York: Seven Stories Press, 2006), 464; Fromm, *Escape from Freedom*, 91. Max Weber has also written on this link between Protestant thought and capitalism; see Max Weber, *The Protestant Ethic and the "Spirit" of Capitalism and Other Writings*, ed. and trans. Peter Baehr and Gordon C. Wells (New York: Penguin Books, 2002).

18. Robert William Fogel and Stanley L. Engerman, *Time on the Cross: The Economics of American Negro Slavery* (1974; reprint, New York: W. W. Norton and Company, 1995), 158–161; Mencken, "Puritanism as a Literary Force," in *The American Intellectual Tradition*, 197.

19. William Graham Sumner, "Sociology," in *The American Intellectual Tradition*, 22; Higham, *Strangers in the Land*, 7–9; Karl Marx and Frederick Engels, *The Communist Manifesto* (New York: International Publishers, 1997), 44.

20. Sumner, "Sociology," in *The American Intellectual Tradtion*, 22–23; Higham, *Strangers in the Land*, 30–32, 54–56, 111; Eric Hobsbawm, *The Age of Capital, 1848–1875* (New York: Vintage Books, 1996), 167–169; Ewen and Ewen, *Typecasting*, 224.

21. Willson, *The Fourth Reader of the United States*, 55–57, 117–118; Elson, *Guardians of Tradition*, 226; Booker T. Washington, *Up From Slavery* (1901; reprint, with a new introduction by Ishmael Reed, New York: Signet Classic, 2000), 64–92; W. E. B. Du Bois, *The Souls of Black Folk* (1903; reprint, with a new introduction by Randall Kenan, New York: Signet Classic, 1995), 79–95; James D. Anderson, *The Education of Blacks in the South, 1860–1935* (Chapel Hill: University of North Carolina Press, 1988), 33–78; Herbert M. Kliebard, *Schooled to Work: Vocationalism and the American Curriculum, 1876–1946* (New York: Teachers College Press, 1999), 26–54; David Nasaw, *Schooled to Order: A Social History of Public Schooling in the United States* (Oxford: Oxford University Press, 1981), 126–145; Woodrow Wilson, "The Ideals of America," in *The American Intellectual Tradition*, 124–130.

22. *Report of the Secretary of the Interior; Being Part of the Message and Documents Communicated to the Two Houses of Congress at the Beginning of the First Session of the Fiftieth Congress*, vol. 2 (Washington, DC: Government Printing Office, 1887), 760–762; *Report of the Secretary of the Interior; Being Part of the Message and Documents Communicated to the Two Houses of Congress at the Beginning of the Second Session of the Forty-Sixth Congress*, vol. 1 (Washington, DC: Government Printing

Office, 1879), 5; David Wallace Adams, *Education for Extinction: American Indians and the Boarding School Experience* (Lawrence: University Press of Kansas, 1995), 16–59.

23. Ronald Takaki, *Strangers from a Different Shore: A History of Asian Americans* (Boston: Little, Brown and Company, 1989), 80–82; Swett, *Public Education in California*, 117; Einar Haugen, *The Norwegian Language in America: A Study of Bilingual Behavior*, vol. 1 (Philadelphia: University of Pennsylvania Press, 1953), 26–27.

24. Steven L. Schlossman, "Is There an American Tradition of Bilingual Education?: German in the Public Elementary Schools, 1840–1919," *American Journal of Education* 91, no. 2 (1983): 165–171; Mary L. Hinsdale, "A Legislative History of the Public School System of the State of Ohio," in *Annual Reports of the Department of the Interior for the Fiscal Year Ended June 30, 1901*, vol. 1 (Washington, DC: Government Printing Office, 1902), 153; Heinz Kloss, *The American Bilingual Tradition* (1977; reprint, with a new introduction by Reynaldo F. Macías and Terrence G. Wiley, McHenry, IL: Center for Applied Linguistics and Delta Systems, 1998), 107.

25. Robert H. Wiebe, *The Search for Order, 1877–1920* (New York: Hill and Wang, 1967), 46; *Thirty-Third Annual Report of the Board of President and Directors of the St. Louis Public Schools for the Year Ending June 30, 1887* (St. Louis: Nixon-Jones Publishing Co., 1888), 15; Dena Lange, "Information Concerning One Hundred Years of Progress in the St. Louis Public Schools," *Public School Messenger* 35, no. 5 (1938): 60; Ramsey, "Let Virtue Be Thy Guide," 143–146; Michael Probstfeld, November 22, 1892, in *News from the Land of Freedom*, 244; Kloss, *The American Bilingual Tradition*, 114; Schlossman, "Is There an American Tradition of Bilingual Education?," 165–175; Selwyn K. Troen, *The Public and the Schools: Shaping the St. Louis System, 1838–1920* (Columbia: University of Missouri Press, 1975), 68–78.

26. Louis Menand, *The Metaphysical Club* (New York: Farrar, Straus and Giroux, 2001), x, 61, 347–358; Oliver Wendell Holmes, Jr., "Natural Law," in *The American Intellectual Tradition*, 138; Marcus, *The Other Victorians*, 284–285; William James, "What Pragmatism Means," in *The American Intellectual Tradition*, 115–117; William James, "The Moral Philosopher and the Moral Life," in *Essays in Pragmatism*, ed. Alburey Castell (New York: Hafner Press, 1948), 65.

27. John Dewey, "Logical Conditions of a Scientific Treatment of Morality," in *John Dewey on Education: Selected Writings*, ed. Reginald D. Archambault (Chicago: University of Chicago Press, 1974), 23–60; Reuben, *The Making of the Modern University*, 133–138; David Starr Jordan, *The Value of Higher Education: An Address to Young People* (Richmond, IN: Daily Palladium Book and Job Printing House, 1888), 16; Paul J. Ramsey, "Building a 'Real' University in the Woodlands of Indiana: The Jordan Administration, 1885–1891," *American Educational History Journal* 31, no. 1 (2004): 20–28; David Starr Jordan, "Nature Study and Moral Culture," *Science* 4, no. 84 (1896): 153.

28. Higham, *Strangers in the Land*, 9–11, 131–157; Winthrop D. Jordan, *The White Man's Burden: Historical Origins of Racism in the United States* (London: Oxford University Press, 1974), 194–204; Takaki, *A Different Mirror*, 84–138, 166–221; Manuel G. Gonzales, *Mexicanos: A History of Mexicans in the United States* (Bloomington: Indiana University Press, 2000), 82–112; Arthur O. Lovejoy, *The Great Chain of Being: A Study of the History of an Idea* (Cambridge, MA: Harvard University Press, 1936), 187–235; Ewen and Ewen, *Typecasting*, 59–68, 279–294; Madison Grant, "The Passing of the Great Race," in *Documents of American Prejudice*, 31–36.

29. Ewen and Ewen, *Typecasting*, 45–68, 101–108, 205–221; Sumner, "Sociology," in *The American Intellectual Tradition*, 20–29; William Graham Sumner, "The Challenge of Facts," in *Documents of American Prejudice*, 168–174; Percy Stickney Grant, "American Ideals and Race Mixture," *North American Review* 195, no. 4 (1912): 513–525; U.S. Immigration Commission, *Statistical Review of Immigration, 1820–1910—Distribution of Immigrants, 1850–1900* (Washington, DC: Government Printing Office, 1911), 12, 393–394; Higham, *Strangers in the Land*, 133–160.

30. Ewen and Ewen, *Typecasting*, 206–221, 235–253; Daniel G. Brinton, *Races and Peoples: Lectures on the Science of Ethnography* (New York: N. D. C. Hodges, 1890), 24–25, 47–48; U.S. Immigration Commission, *Dictionary of Races or Peoples* (Washington: GPO, 1911), 1–10, 54–57, 64–68, 81–85; Higham, *Strangers in the Land*, 131–157; Henry F. May, *The End of American Innocence: A Study of the First Years of Our Own Time, 1912–1917* (New York: Alfred A. Knopf, 1959), 37–39; Michael B. Katz, *The Irony of Early School Reform: Educational Innovation in Mid-Nineteenth Century Massachusetts* (New York: Teachers College Press, 2001), 170–180; Prescott F. Hall, "Italian Immigration," *North American Review* 163, no. 2 (1896): 252–254; Prescott F. Hall, "The Future of American Ideals," *North American Review* 195, no. 1 (1912): 94–102.

31. Lester Frank Ward, "Mind as a Social Factor," in *The American Intellectual Tradition*, 35; Richard Hofstadter, *Social Darwinism in American Thought* (1944; reprint, with a new introduction by Eric Foner, Boston: Beacon Press, 1992), 67–84, 143–169; Joseph M. Hawes, *Children in Urban Society: Juvenile Delinquency in Nineteenth-Century America* (New York: Oxford University Press, 1971), 197–208; Paul J. Ramsey, "Wrestling with Modernity: Philanthropy and the Children's Aid Society in Progressive-Era New York City," *New York History* 88, no. 2 (2007): 153–174; Grant, "American Ideals and Race Mixture," 513–525.

32. Israel Zangwill, "America Is a Great Melting Pot," in *Immigration*, 46; Gary Gerstle, *American Crucible: Race and Nation in the Twentieth Century* (Princeton: Princeton University Press, 2001), 3–11; Wiebe, *The Segmented Society*, 68; Stephan F. Brumberg, *Going to America, Going to School: The Jewish Immigrant Public School Encounter in Turn-of-the-Century New York City* (New York: Praeger, 1986), 23; Elson, *Guardians of Tradition*, 54; Adele Marie Shaw, "The True Character of the New York Public Schools,"

in *Woman's "True" Profession: Voices from the History of Teaching*, ed. Nancy Hoffman (Old Westbury, NY: Feminist Press, 1981), 229, 232.

33. Paula S. Fass, *Outside In: Minorities and the Transformation of American Education* (New York: Oxford University Press, 1989), 22–30; Robert A. Carlson, *The Quest for Conformity: Americanization through Education* (New York: John Wiley and Sons, Inc., 1975), 83–93; John F. McClymer, "The Americanization Movement and the Education of the Foreign-Born Adult, 1914–25," in *American Education and the European Immigrant*, 96–116; Jeffrey Mirel, "Civic Education and Changing Definitions of American Identity, 1900–1950," *Educational Review* 54, no. 2 (2002): 143–146; William L. Ettinger, *Ten Addresses Delivered before Associate and District Superintendents of the New York City Schools and Other Professional Bodies* (n.p., n.d.), 24–25, 29.

34. Józef Miaso, *The History of the Education of Polish Immigrants in the United States*, trans. Ludwik Krzyżanowski (New York: Kosciuszko Foundation, 1977), 105–126; Carlos Kevin Blanton, *The Strange Career of Bilingual Education in Texas, 1836–1981* (College Station: Texas A&M University Press, 2004), 37–41; William J. Reese, *Power and the Promise of School Reform: Grassroots Movements during the Progressive Era* (New York: Teachers College Press, 2002), 116–118; Milwaukee Public Schools, *Fifty-Seventh Annual Report of the Board of School Directors of the City of Milwaukee for the Year Ending June 30, 1916* (Milwaukee: Radtke Bros. and Kortsh Co., n.d.), 61–63.

35. Selma Cantor Berrol, *Immigrants at School: New York City, 1898–1914* (New York: Arno Press, 1978), 50–51, 210–228; Selma Berrol, "Public Schools and Immigrants: The New York City Experience," in *American Education and the European Immigrant*, 36; Brumberg, *Going to America, Going to School*, 74–75, 86–87; Department of Education, the City of New York, *Fourth Annual Report of the City Superintendent of Schools for the Year Ending July 31, 1902* (New York: n.p., n.d.), 177–181; Ramsey, "Let Virtue Be Thy Guide," 140–146.

36. May, *The End of American Innocence*, 3–9, 347–350; Hall, "The Future of American Ideals," 94–102; Ewen and Ewen, *Typecasting*, 282–294; Edward Alsworth Ross, "The New Immigrants Harm American Society," in *Immigration*, 128–138; Jacob A. Riis, *How the Other Half Lives* (1890; reprint, with an introduction by Francesco Cordasco, New York: Garrett Press, Inc., 1970), 48–53, 104–113; Christopher Lasch, *Haven in a Heartless World: The Family Besieged* (New York: Basic Books, Inc., 1977), 12–21; Ramsey, "Wrestling with Modernity," 153–174.

6   Progressivism, Science, and the Remodeling of the Boardinghouse

1. Theodore Dreiser, *A Hoosier Holiday* (1916; reprint, with a new introduction by Douglas Brinkley, Bloomington: Indiana University Press, 1998), 171; Theodore Roosevelt, "The Nation Should Assume Power of Regulation

over All Corporations," in *Our Nation's Archive: The History of the United States in Documents*, ed. Erik Bruun and Jay Crosby (New York: Tess Press, 1999), 577; Robert H. Wiebe, *The Search for Order, 1877–1920* (New York: Hill and Wang, 1967), xviii–xiv, 189–195; Theodore Roosevelt, "The New Nationalism," in *The Progressive Movement, 1900–1915*, ed. Richard Hofstadter (New York: Touchstone, 1986), 125.

2. William J. Reese, *Power and the Promise of School Reform: Grassroots Movements during the Progressive Era* (New York: Teachers College Press, 2002), 16–26; Wiebe, *The Search for Order*, 7–27, 44–66; Robert M. La Follette, "Railroad Regulation," in *The Progressive Movement*, 148; Sigmund Freud, *Civilization and Its Discontents*, trans. James Strachey (1930; reprint, with an introduction by Peter Gay, New York: W. W. Norton and Company, 1989), 44.

3. Wiebe, *The Search for Order*, 50–56; John Fiske, "Letter to William Lloyd Garrison," in *Documents of American Prejudice: An Anthology of Writings on Race from Thomas Jefferson to David Duke*, ed. S. T. Joshi (New York: Basic Books, 1999), 513–515.

4. Morton Keller, *Regulating a New Society: Public Policy and Social Change in America, 1900–1933* (Cambridge, MA: Harvard University Press, 1994), 4; Dreiser, *A Hoosier Holiday*, 172; Richard Hofstadter, ed., *The Progressive Movement, 1900–1915* (New York: Touchstone, 1986), 4; Wiebe, *The Search for Order*, 91–92, 142–143; Eugene Debs, "While There Is a Lower Class, I Am in It," in *Our Nation's Archive*, 632–633; Theodore Debs, *Sidelights: Incidents in the Life of Eugene V. Debs* (Terre Haute, IN: Moore-Langen Printing Co., n.d.), 15–18; Walter Rauschenbusch, "The Social Role of Christianity," in *The Progressive Movement*, 79–83; Samuel Gompers, "The Needs of Labor," in *The Progressive Movement*, 101–103; Edward A. Ross, "The Criminaloid Type," in *The Progressive Movement*, 73–78.

5. Richard Hofstadter, *The Age of Reform: From Bryan to F.D.R.* (New York: Vintage Books, 1955), 60–70; Wiebe, *The Search for Order*, 8–9, 84–85, 98–100; Michael Probstfeld, November 22, 1892, in *News from the Land of Freedom: German Immigrants Write Home*, ed. Walter D. Kamphoefner, Wolfgang Helbich, and Ulrike Sommer, trans. Susan Carter Vogel (Ithaca: Cornell University Press, 1991), 244; Michael Probstfeld, June 22, 1893, in *News from the Land of Freedom*, 248–249; Michael Probstfeld, January 20, 1894, in *News from the Land of Freedom*, 251; Horatio Alger, Jr., *Ragged Dick: Or, Street Life in New York with the Boot-blacks* (1868; reprint, with an introduction by Alan Trachtenberg, New York: Signet Classic, 1990), 79; Reese, *Power and the Promise of School Reform*, 24.

6. Wiebe, *The Search for Order*, 111–132, 147; Judith Sealander, "Curing Evils at Their Source: The Arrival of Scientific Giving," in *Charity, Philanthropy, and Civility in American History*, ed. Lawrence J. Friedman and Mark D. McGarvie (New York: Cambridge University Press, 2003), 217–239; Bruce Mazlish, *The Uncertain Sciences* (New Haven: Yale University Press, 1998), 59–66; Sinclair Lewis, *Babbitt* (1922; reprint, Mineola, NY: Dover Publications, Inc., 2003), 53, 64, 77.

7. David Riesman, in collaboration with Reuel Denny and Nathan Glazer, *The Lonely Crowd: A Study of the Changing American Character* (New Haven: Yale University Press, 1950), 15, 22; B. Edward McClellan, *Moral Education in America: Schools and the Shaping of Character from the Colonial Times to the Present* (New York: Teachers College Press, 1999), 46–47; Paul J. Ramsey, " 'Let Virtue Be Thy Guide, and Truth Thy Beacon-Light": Moral and Civic Transformation in Indianapolis's Public Schools," in *Civic and Moral Learning in America*, ed. Donald Warren and John J. Patrick (New York: Palgrave Macmillan, 2006), 135–151; Wiebe, *The Search for Order*, 145–155; Lewis, *Babbitt*, 38, 54, 59, 87; Dreiser, *A Hoosier Holiday*, 106.

8. Wiebe, *The Search for Order*, 166; Lester Frank Ward, "Mind as a Social Factor," in *The American Intellectual Tradition: A Sourcebook*, ed. David A. Hollinger and Charles Capper, vol. 2, 4th ed. (New York: Oxford University Press, 2001), 38; Jacob A. Riis, *How the Other Half Lives* (1890; reprint, with an introduction by Francesco Cordasco, New York: Garrett Press, Inc., 1970), 32–46, 121–128; Upton Sinclair, *The Jungle* (1906; reprint, with an introduction by Ronald Gottesman, New York: Penguin Books, 1986), 9, 34–35, 39, 84, 107, 121, 162–164; S. S. McClure, "A Trend of the Times," in *The Progressive Movement*, 16–17; Ida M. Tarbell, "The Methods of the Standard Oil Company," in *The Progressive Movement*, 20–27.

9. Wiebe, *The Search for Order*, 166–195; Hofstadter, *The Age of Reform*, 215–271; Sinclair, *The Jungle*, 199; Stephen Skowronek, *Building a New American State: The Expansion of National Administrative Capacities, 1877–1920* (New York: Cambridge University Press, 1982), 47–84, 177–211; Keller, *Regulating a New Society*, 205–209.

10. Steven Mintz, *Huck's Raft: A History of American Childhood* (Cambridge, MA: Belknap Press, 2004), 154–184; David Nasaw, *Children of the City: At Work and at Play* (Garden City, NY: Anchor Press/Doubleday, 1985), 17–87, 91–94; LeRoy Ashby, *Saving the Waifs: Reformers and Dependent Children, 1890–1917* (Philadelphia: Temple University Press, 1984), 3–37; Paul J. Ramsey, "Wrestling with Modernity: Philanthropy and the Children's Aid Society in Progressive-Era New York City," *New York History* 88, no. 2 (2007): 153–174; Saul Bellow, *The Adventures of Augie March* (New York: Penguin Books, 1996), 14, 287; Sinclair, *The Jungle*, 35–37; Leonard Covello, with Guido D'Agostino, *The Heart Is the Teacher* (New York: McGraw-Hill, 1958), 28–29; Jane Addams, *Twenty Years at Hull-House, with Autobiographical Notes* (1919; reprint, with a foreword by Henry Steele Commager, New York: Signet Classic, 1961), 181; Mary Antin, *The Promised Land*, 2nd ed. (Princeton: Princeton University Press, 1969), 100; Willa Cather, *My Ántonia* (1918; reprint, Boston: Houghton Mifflin, 1954), 123; Elliot West, *Growing Up with the Country: Childhood on the Far Western Frontier* (Albuquerque: University of New Mexico Press, 1989), 73–98; H. L. Bliss, "Census Statistics of Child Labor," in *Children and Youth in America: A Documentary History*, vol. 2, ed. Robert H. Bremner (Cambridge, MA: Harvard University Press, 1971), 605.

11. Reese, *Power and the Promise of School Reform*, 27–55; Sara M. Evans, *Born for Liberty: A History of Women in America* (New York: Free

Press, 1989), 145–173; Ellen Wiley Todd, *The "New Woman" Revised: Painting and Gender Politics on Fourteenth Street* (Berkley: University of California Press, 1993), 1–24; Charlotte Perkins Gilman, "Selection from *Women and Economics*," in *The American Intellectual Tradition*, 46–47; Jane Addams, "The College Woman and the Family Claim," in *On Education*, ed. Ellen Condliffe Lagemann (New Brunswick: Transaction Publishers, 2002), 64–73; Sarah Henry Lederman, "Philanthropy and Social Case Work: Mary E. Richmond and the Russell Sage Foundation, 1909–1928," in *Women and Philanthropy in Education*, ed. Andrea Walton (Bloomington: Indiana University Press, 2005), 60–80; Michael B. Katz, *In the Shadow of the Poorhouse: A Social History of Welfare in America*, 10th anniversary ed. (New York: Basic Books, 1996), 117–139; Emily D. Cahan, "Toward a Socially Relevant Science: Notes on the History of Child Development Research," in *When Science Encounters the Child: Education, Parenting, and Child Welfare in 20th-Century America*, ed. Barbara Beatty, Emily D. Cahan, and Julia Grant (New York Teachers College Press, 2006), 18–21; Mrs. Max West, *Infant Care* (Washington, DC: Government Printing Office, 1914); U.S. Children's Bureau, *Baby-Saving Campaigns: A Preliminary Report on What American Cities Are Doing to Prevent Infant Mortality* (Washington, DC: Government Printing Office, 1913).

12. David B. Tyack, *The One Best System: A History of American Urban Education* (Cambridge, MA: Harvard University Press, 1974), 39–59, 126–147; Reese, *Power and Promise of School Reform*, 79–99; James Jackson Storrow, "Son of New England," in *The School in the United States: A Documentary History*, ed. James W. Fraser (Boston: McGraw-Hill, 2001), 183–187.

13. Reese, *Power and Promise of School Reform*, 28–35, 79–130; Ella Flagg Young, "Isolation in the School," in *The School in the United States*, 191–194; Tyack, *The One Best System*, 126–176; Jeffrey Mirel, *The Rise and Fall of an Urban School System: Detroit, 1907–81* (Ann Arbor: University of Michigan Press, 1993), 27–31; Storrow, "Son of New England," in *The School in the United States*, 184–185.

14. Tyack, *The One Best System*, 126–176; Skowronek, *Building a New American State*, 177–211; Storrow, "Son of New England," in *The School in the United States*, 183–187; Reese, *Power and the Promise of School Reform*, 100–130; Margaret Haley, "Why Teachers Should Organize," in *The School in the United States*, 187–191.

15. Ellwood P. Cubberley, *Public Education in the United States: A Study and Interpretation of American Educational History* (Boston: Houghton Mifflin Company, 1919), xiii; William J. Reese, *America's Public Schools: From the Common School to "No Child Left Behind"* (Baltimore: Johns Hopkins University Press, 2005), 142–146; Herbert M. Kliebard, *The Struggle for the American Curriculum, 1893–1958*, 3rd ed. (New York: RoutledgeFalmer, 2004), 8–11; David L. Angus and Jeffrey E. Mirel, "Presidents Professors, and Lay Boards of Education: The Struggle for Influence over the American High School, 1860–1910," in *A Faithful Mirror: Reflections on the College*

Board and Education in America, ed. Michael C. Johanek (New York: College Board, 2001), 3–42.

16. Kliebard, *The Struggle for the American Curriculum*, 7, 36–44, 51–75, 92, 130–150, 158–161; Lawrence A. Cremin, *The Transformation of the School: Progressivism in American Education, 1876–1957* (New York: Vintage Books, 1964), 100–105, 111–126, 305–306; Cahan, "Toward a Socially Relevant Science," 17–22; Francis W. Parker, "New Methods of Instruction in the Public Schools of Quincy, Massachusetts, 1876," in *Children and Youth in America*, vol. 2, 1131–1132; Paula S. Fass, *Children of a New World: Society, Culture, and Globalization* (New York: New York University Press, 2007), 52–53; John Dewey, *Democracy and Education: An Introduction to the Philosophy of Education* (1916; reprint, New York: Free Press, 1997), 51; John Dewey, *Experience and Education* (1938; reprint, New York: Touchstone, 1997), 17–23.

17. Kliebard, *The Struggle for the American Curriculum*, 76–104, 106; Herbert M. Kliebard, *Schooled to Work: Vocationalism and the American Curriculum, 1876–1946* (New York: Teachers College Press, 1999), 153–162; Paul W. Horn, *Survey of the City Schools of El Paso Texas* (El Paso: Department of Printing of the City Schools, El Paso, Texas, 1922), 34; J. F. Bobbitt, *The San Antonio Public School System: A Survey* (San Antonio: San Antonio School Board, 1915), 124.

18. Lewis M. Terman, "National Intelligence Tests," in *The School in the United States*, 207–213; Fass, *Children of a New World*, 49–73.

19. David L. Angus and Jeffrey E. Mirel, *The Failed Promise of the American High School, 1890–1995* (New York: Teachers College Press, 1999), 14–16, 32–38; William G. Wraga, "A Progressive Legacy Squandered: The *Cardinal Principles* Report Reconsidered," *History of Education Quarterly* 41, no. 4 (2001): 494–519; Reese, *America's Public Schools*, 191–193, 254; Larry Cuban, *How Teachers Taught: Constancy and Change in American Classrooms, 1880–1990*, 2nd ed. (New York: Teachers College Press, 1993), 23–75.

20. Reese, *America's Public Schools*, 118–119; Selma C. Berrol, *Julia Richman: A Notable Woman* (Philadelphia: Balch Institute Press, 1993), 61–72; "New York Compulsory Education Law, 1909" in *Children and Youth in America*, vol. 2, 1424–1425; U.S. Children's Bureau, "Outline of Legislation, 1830–1929," in *Children and Youth in America*, 666–668; Florence Kelley, "Objective Tests of the Enforcement of State Laws, 1907," in *Children and Youth in America*, 677–679; N. C. Dougherty, "Recent Legislation upon Compulsory Education in Illinois and Wisconsin," in *Journal of Proceedings and Addresses, National Education Association* (New York: National Education Association, 1891), 393–401; Fass, *Children of a New World*, 23–38; Ramsey, "Let Virtue Be Thy Guide," 135–151; *Fourth Biennial Report of the Superintendent of Public Instruction to the Governor of North Dakota, for the Two Years Ending June 30, 1896* (Jamestown, N.D.: Alert, State Printers and Binders, 1896), 16.

21. Fass, *Children of a New World*, 30–38; Dreiser, *Hoosier Holiday*, 66; Franz Joseph Löwen, April 29, 1888, in *News from the Land of Freedom*, 200;

Heinrich Möller, February 18, 1886, in *News from the Land of Freedom*, 218; Michael Probstfeld, December 13, 1890, in *News from the Land of Freedom*, 238; Einar Haugen, *The Norwegian Language in America: A Study of Bilingual Behavior*, vol. 1 (Philadelphia: University of Pennsylvania Press, 1953), 52–63; U.S. Immigration Commission, *The Children of Immigrants in Schools*, vol. 1 (Washington, DC: Government Printing Office, 1911), 83–86, 97–99.

22. Fass, *Children of a New World*, 30–38; Nasaw, *Children of the City*, 17–38, 115–129; Antin, *The Promised Land*, 184–185; Johann Witten, February 3, 1908, in *News from the Land of Freedom*, 271; Addams, *Twenty Years at Hull House*, 172, 180–181; Dreiser, *Hoosier Holiday*, 68–69.

23. Tyack, *The One Best System*, 39–59, 108–109; Lewis, *Babbitt*, 53; L. Viereck, "German Instruction in American Schools," in *Annual Reports of the Department of the Interior for the Fiscal Year Ended June 30, 1901*, vol. 1 (Washington, DC: Government Printing Office, 1902), 640; Heinz Kloss, *The American Bilingual Tradition* (1977; reprint, with a new introduction by Reynaldo F. Macías and Terrence G. Wiley, McHenry, IL: Center for Applied Linguistics and Delta Systems Co., Inc., 1998), 114; Linda L. Sommerfeld, "An Historical Descriptive Study of the Circumstances That Led to the Elimination of German from the Cleveland Schools, 1860–1918 (PhD diss., Kent State University, 1986), 121–124; *Biennial Report, North Dakota, 1896*, 15–16, 283–284; Carlos Kevin Blanton, *The Strange Career of Bilingual Education in Texas, 1836–1981* (College Station: Texas A&M University Press, 2004), 29–30, 52–55; W. F. Doughty and R. B. Binnion, *School Laws of Texas, 1913* (Austin: Von [Boe]ckmann-Jones Company, Printers, n.d.), 82–83; California Council of Education, *First Report of the Committee on the Course of Study for Elementary Schools* (Oakland: n.p., 1896), 26–27; Ramsey, "Let Virtue Be Thy Guide," 143–146.

24. Herbert Adolphus Miller, *Cleveland Educational Survey: The School and the Immigrant* (Cleveland: Survey Committee of the Cleveland Foundation, 1916), 11–31, 72–84; Carlos Kevin Blanton, "The Rise of English-Only Pedagogy: Immigrant Children, Progressive Education, and Language Policy in the United States, 1900–1930," in *When Science Encounter the Child*, 56–76.

25. Jessie Rosenfeld, "Special Classes in the Public Schools of New York," *Education* 27, no. 2 (1906): 92–100; Department of Education, The City of New York, *Fourth Annual Report of the City Superintendent of Schools for the Year Ending July 31, 1902* (New York: n.p., n.d.), 177–181; U.S. Bureau of Education, *Education of the Immigrant* (Washington, DC: Government Printing Office, 1913), 21; Selma Cantor Berrol, *Immigrants at School: New York City, 1898–1914* (New York: Arno Press, 1978), 196–228; Berrol, *Julia Richman*, 61, 72–77; Stephan F. Brumberg, *Going to America, Going to School: The Jewish Immigrant Public School Encounter in Turn-of-the-Century New York City* (New York: Praeger, 1986), 60–70.

26. Miller, *Cleveland Education Survey*, 75–76; Charles F. Kroeh, "Methods of Teaching Modern Languages," *Transactions and Proceedings of the Modern Language Association of America* 3 (1887): 169–185; Charles F. Kroeh, H. C. G. von Jagemann, L. A. Staeger, O. Seidensticker, Paul F. Rohrbacher,

and C. Sprague Smith, "Methods of Teaching Modern Languages: Discussion," *Transactions and Proceedings of the Modern Language Association of America* 3 (1887): xxxiii–xxxvi; Blanton, "The Rise of English-Only Pedagogy," 56–76.

27. Charles W. Eliot, "Address of Welcome," *Publications of the Modern Language Association of America* 5, no. 1 (1890): 2; Blanton, "The Rise of English-Only Pedagogy," 56–76; Blanton, *Strange Career*, 74–91; Frank V. Thompson, *Schooling of the Immigrant* (New York: Harper and Brothers Publishers, 1920), 164–213; U.S. Bureau of Education, *Education of the Immigrant*, 18–42; *Annual Report, New York City, 1902*, 177–181; Miller, *Cleveland Education Survey*, 75–77.

28. Michael R. Olneck and Marvin Lazerson, "The School Achievement of Immigrant Children: 1900–1930," *History of Education Quarterly* 14, no. 4 (1974): 453–482; David Hogan, "Education and the Making of the Chicago Working Class, 1880–1930," *History of Education Quarterly* 18, no. 3 (1978): 227–270; Selma Cantor Berrol, *Growing Up American: Immigrant Children in America, Then and Now* (New York: Twayne Publishers, 1995), 34–38; Joel Perlmann, *Ethnic Differences: Schooling and School Structure among the Irish, Italians, Jews, and Blacks in an American City* (Cambridge: Cambridge University Press, 1988), 83–121; Sinclair, *The Jungle*, 54–89.

29. U.S. Immigration Commission, *The Children of Immigrants in Schools*, vol. 1, 144–153; Miller, *Cleveland Education Survey*, 31–33, 37–53; U.S. Bureau of Education, *A Survey of Education in Hawaii* (Washington, DC: Government Printing Office, 1920), 107–143; Ronald Takaki, *Strangers from a Different Shore: A History of Asian Americans* (Boston: Little, Brown and Company, 1989), 132–176; Eileen H. Tamura, "The English-Only Effort, the Anti-Japanese Campaign, and Language Acquisition in the Education of Japanese Americans in Hawaii, 1915–40," *History of Education Quarterly* 33, no. 1 (1993): 37–58; Louisiana State Department of Education, *Public School Statistics: Session of 1914–'15: Part II of the Biennial Report of 1913–'14 and 1914–'15* (Baton Rouge: Ramires-Jones Printing Company, 1915), 72; T. Lynn Smith and Vernon J. Parenton, "Acculturation among the Louisiana French," *American Journal of Sociology* 44, no. 3 (1938): 355–364; Blanton, *Strange Career*, 28–30, 42–55; Steven L. Schlossman, "Is There an American Tradition of Bilingual Education?: German in the Public Elementary Schools, 1840–1919," *American Journal of Education* 90, no. 2 (1983): 139–186; *Fifteenth Biennial Report of the State Superintendent of Public Instruction to the Governor of the State of Nebraska, Dec. 1, 1898* (Lincoln: Jacob North and Company, 1898), 187–188, 206; *Biennial Report, North Dakota, 1896*, 283–284; Charles Hart Handschin, *The Teaching of Modern Languages in the United States* (Washington, DC: Government Printing Office, 1913), 28–29, 67–74; Rosenfeld, "Special Classes in the Public Schools of New York," 99.

30. *Seventeenth and Eighteenth Annual Reports of the Territorial Superintendent of Public Instruction to the Governor of New Mexico for the Years 1907–1908* (Santa Fe: New Mexican Printing Company, 1909), 11, 17–18; John M. Nieto-Phillips, *The Language of Blood: The Making*

of Spanish-American Identity in New Mexico, 1880s-1930s (Albuquerque: University of New Mexico Press, 2004), 198–200.

31. *Nineteenth and Twentieth Annual Reports of the Territorial Superintendent of Public Instruction to the Governor of New Mexico for the Years 1909–1910* (Santa Fe: New Mexican Printing Company, 1911), 26–29, 35–38, 48–49, 62, 88–93, 144; Nieto-Phillips, *The Language of Blood*, 200.

32. Viereck, "German Instruction in American Schools," 640; Miller, *Cleveland Education Survey*, 41; Sommerfeld, "Elimination of German from the Cleveland School," 121–124; Public Schools of Cincinnati, *Eighty-Second Annual Report of the Public Schools of Cincinnati for the School Year Ending August 31, 1911* (n.p., n.d.), 41; Carolyn R. Toth, *German-English Bilingual Schools in America: The Cincinnati Tradition in Historical Context* (New York: Peter Lang, 1990), 61, 63–69.

33. Indianapolis Public Schools, *Annual Report of the Secretary, Business Director, Superintendent of Schools, and the Librarian* (n.p., 1902), 87–92; Indianapolis Public Schools, *Annual Report of the Secretary, Business Director, Superintendent of Schools and the Librarian* (n.p., [1909?]), 152–154; Ramsey, "Let Virtue Be Thy Guide," 139–146; Indianapolis Public Schools, *Annual Report: Business Director, Superintendent of Schools and Librarian* (n.p., 1916), 130.

34. Sommerfeld, "Elimination of German from the Cleveland Schools," 92–167; M. D. Learned, "German in the Public Schools," *German American Annals*, vol. 2 (1913):100–106; Viereck, "German Instruction in American Schools," 576–580, 646.

35. Milwaukee Public Schools, *Fifty-Seventh Annual Report of the Board of School Directors of the City of Milwaukee for the Year Ending June 30, 1916* (Milwaukee: Radtke Bros. and Kortsch Co., n.d.), 21, 61–63; Reese, *Power and the Promise of School Reform*, 109–123; Jane Addams, "The Public School and the Immigrant Child," in *On Education*, 136–142.

7   "If You Can't Fight over There, Fight over Here": World War I and the Partial Destruction of the Boardinghouse

1. David M. Kennedy, *Over Here: The First World War and American Society*, 25th anniversary ed. (New York: Oxford University Press, 2004), 3–14; "Comment of This Morning's Newspapers on President Wilson's Announcement," *New York Times*, February 4, 1917, 5.

2. Kennedy, *Over Here*, 3–14; "Berlin Sees a Wilson Trick," *New York Times*, March 6, 1917, 1, 4; "The Zimmerman Note," in *Our Nation's Archive: The History of the United States in Documents*, ed. Erik Bruun and Jay Crosby (New York: Tess Press, 1999), 622. "Wilson's Declaration of War against Germany," in *Our Nation's Archive*, 623–627; "For Freedom and Civilization," *New York Times*, April 3, 1917, 12.

3. Paul J. Ramsey, "The War against German-American Culture: The Removal of German-Language Instruction from the Indianapolis Schools, 1917–1919," *Indiana Magazine of History* 98, no. 4 (2002): 285–303; George M. Cohan, "Over There," in *Our Nation's Archive*, 627; Theodore Roosevelt, "One Flag, One Language," in *Language Loyalties: A Source Book on the Official English Controversy*, ed. James Crawford (Chicago: University of Chicago Press, 1992), 84–85; L. N. Hines, ed., *Proceedings and Papers of the Indiana State Teachers' Association, October 31, and November 1, 2, 3, 1917* (n.p., n.d.), 350.

4. Kennedy, *Over Here*, 45–69; Robert H. Wiebe, *The Search for Order, 1877–1920* (New York: Hill and Wang, 1967), xiii, 224–255; John Dos Passos, *1919* (New York: Harcourt, Brace and Company, 1932), 16; Louis-Ferdinand Céline, *Journey to the End of the Night*, trans. Ralph Manheim (New York: New Directions, 2006), 6, 9, 13.

5. Carl Wittke, *The German-Language Press in America* (University of Kentucky Press,1957), 237–240; Ludwig Dilger, December 6, 1914, in *News from the Land of Freedom: German Immigrants Write Home*, ed. Walter D. Kamphoefner, Wolfgang Helbich, and Ulrike Sommer, trans. Susan Carter Vogel (Ithaca: Cornell University Press, 1991), 503; Kennedy, *Over Here*, 59–66; Lewis Paul Todd, *Wartime Relations of the Federal Government and the Public Schools, 1917–1918* (New York: Teachers College, 1945), 13–39; Dos Passos, *1919*, 105; U.S. Bureau of Education, *Education in Patriotism: A Synopsis of the Agencies at Work* (Teachers' Leaflet, no. 2, 1918), 1–3; Frederick C. Luebke, *Bonds of Loyalty: German-Americans and World War I* (DeKalb, IL: Northern Illinois University Press, 1974), 225–259; Erik Kirschbaum, *The Eradication of German Culture in the United States: 1917–1918* (Stuttgart: Hans-Dieter Heinz, 1986), 45–46.

6. Kennedy, *Over Here*, 67–68, 81–83; Wittke, *The German-Language Press*, 245–258; Ramsey, "The War against German-American Culture," 285–303; "Wilson's Declaration of War," in *Our Nation's Archive*, 626; Ludwig Dilger, November 29, 1928, in *News from the Land of Freedom*, 511; Luebke, *Bonds of Loyalty*, 3–10; Colin Ross, *Unser Amerika: Der deutsche Anteil an den Vereinigten Staaten* (Leipzig: F. A. Brockhaus, 1936), 316; Frederick Franklin Schrader, *Handbook: Political, Statistical and Sociological for German Americans and All Other Americans Who Have Not Forgotten the History and Traditions of Their Country and Who Believe in the Principles of Washington, Jefferson and Lincoln* (New York: n.p., 1916–1917), 2; "Esther Vaagen," *North Dakota History* 44, no. 4 (1977): 16; Linda L. Sommerfeld, "An Historical Descriptive Study of the Circumstances That Led to the Elimination of German from the Cleveland Schools, 1860–1918" (PhD diss., Kent State University, 1986), 40–50; Noam Chomsky, *Media Control: The Spectacular Achievements of Propaganda*, 2nd ed. (New York: Seven Stories Press, 2002), 11–12.

7. Hines, *Proceedings and Papers of the Indiana State Teachers' Association, 1917*, 343–344; Kennedy, *Over Here*, 4–12; Wittke, *The German-Language Press*, 240; Kirschbaum, *The Eradication of German Culture*, 46–54; Gustavus Ohlinger, *The German Conspiracy in Education* (New York: George

H. Doran Company, [1919?]), 14–15; Kaiser Wilhelm II, "Das neue Reich," in *Stimmen eines Jahrhunderts 1888–1990: Deutsche Autobiographien, Tagebücher, Bilder und Briefe*, ed. Andreas Lixl-Purcell (Fort Worth: Holt, Rinehart and Winston, Inc., 1990), 49–51; Walter Roy, "Brief an die Eltern, 24. April 1915," in *Stimmen eines Jahrhunderts*, 80–81.

8. Kirschbaum, *The Eradication of German Culture*, 117–135; Ramsey, "The War against German-American Culture," 296–297; Gustavus Ohlinger, *The German Conspiracy in Education*, 9, 18–19; Gustavus Ohlinger, *Their True Faith and Allegiance* (New York: Macmillan Company, 1916), vii–xix; Wittke, *The German-Language Press*, 246–247; David Jayne Hill, "Dual Citizenship in the German Imperial and State Citizenship Law," *American Journal of International Law* 12 (April 1918): 356–363.

9. Kennedy, *Over Here*, 67–88; Dos Passos, *1919*, 245–246; "Wilson's Declaration of War," in *Our Nation's Archive*, 626; Howard Zinn, *A People's History of the United States, 1492–Present* (New York: Perennial Classics, 2001), 329–342, 364–373; Eugene Debs, "While There Is a Lower Class, I Am in It," in *Our Nation's Archive*, 632–633.

10. Bureau of Education, *Education in Patriotism*, 2; Kennedy, *Over Here*, 50–53, 65–66, 75–78; Arlow W. Andersen, *Rough Road to Glory: The Norwegian-American Press Speaks Out on Public Affairs, 1875–1925* (Philadelphia: Balch Institute Press, 1990), 127–146; Dos Passos, *1919*, 105; Ramsey, "The War against German-American Culture," 285–303; Wittke, *The German-Language Press*, 259–261; David W. Detjen, *The Germans in Missouri, 1900–1918: Prohibition, Neutrality, and Assimilation* (Columbia: University of Missouri Press, 1985), 138–160; "To Our Readers," *Telegraph and Tribüne*, May 31, 1918, 1; Frederick C. Luebke, "The German-American Alliance in Nebraska, 1910–1917," *Nebraska History* 49, no. 2 (1968): 165–185; LaVern J. Rippley, *The German-Americans* (Boston: Twayne Publishers, 1976), 191–192.

11. Bureau of Education, *Education in Patriotism*, 1–10; Todd, *Wartime Relations of the Federal Government and the Public Schools*, 17–23; *Fifteenth Annual Report of the Belleville Public Schools for the School Year Ending June 22, 1888, and Various Supplementary Documents, Exhibiting the Conditions of the Schools* (Belleville: Geo. Semmelroth and Co., n.d.), 14–19; *Forty-First Report of the Belleville Public Schools, District No. 118, for the School Years Ending June 22, 1917 and June 21, 1918* (n.p., 1918), 17.

12. Charles A. Coulomb, Armand J. Gerson, and Albert E. McKinley, *Outline of an Emergency Course of Instruction on the War* (Washington, DC: Government Printing Office, 1918), 5–19; Ohlinger, *The German Conspiracy in Education*, 112–113; Hines, *Proceedings and Papers of the Indiana State Teachers' Association, 1917*, 337; Ramsey, "The War against German-American Culture," 297.

13. U.S. Bureau of Education, *Government Policies Involving the Schools in War Time* (Teachers' Leaflet, no. 3, 1918), 1–6; Todd, *Wartime Relations of the Federal Government and the Public Schools*, 73, 96–103, 158–167; Kennedy, *Over Here*, 54; Coulomb, Gerson, and McKinley, *Outline of an*

Emergency Course of Instruction, 6–7; Sylvia Plath, *The Bell Jar* (New York: Perennial Classics, 1999), 33.

14. Edward George Hartmann, *The Movement to Americanize the Immigrant* (New York: AMS Press, Inc., 1967), 13–37, 164–215; Howard C. Hill, "The Americanization Movement," *The American Journal of Sociology* 24, no. 6 (1919): 609–642; Edward Hale Bierstadt, *Aspects of Americanization* (Cincinnati: Stewart Kidd Company, 1922), 26, 93, 98; Sinclair Lewis, *Babbitt* (1922; reprint, Mineola, NY: Dover Publications, Inc., 2003), 297–298.

15. Hartmann, *The Movement to Americanize the Immigrant*, 165–215; Robert A. Carlson, *The Quest for Conformity: Americanization through Education* (New York: John Wiley and Sons, Inc., 1975), 107–131; Hill, "The Americanization Movement," 609–642.

16. Ramsey, "The War against German-American Culture," 285–288, 298–299; Ohlinger, *The German Conspiracy in American Education*, 11–16; Todd, *Wartime Relations of the Federal Government and the Public Schools*, 74; Bierstadt, *Aspects of Americanization*, 15–16, 19; Hill, "The Americanization Movement," 631–632.

17. Ramsey, "The War against German-American Culture," 285–288; Ohlinger, *The German Conspiracy in American Education*, 27–35, 105–106; Clifford Wilcox, "World War I and the Attack on Professors of German at the University of Michigan," *History of Education Quarterly* 33, no. 1 (1993): 59–84.

18. Frederick C. Luebke, *Germans in the New World: Essays in the History of Immigration* (Urbana: University of Illinois Press, 1990), 36–38; Lawrence A. Wilkins, "Spanish as a Substitute for German for Training and Culture" *Hispania* 1, no. 4 (1918): 205–208, 220–221; Ramsey, "The War against German-American Culture," 298–301; *Laws of the State of Indiana Passed at the Seventy-First Regular Session of the General Assembly* (Indianapolis: Wm. B. Burford, 1919), 823.

19. Johann Witten, September 11, 1919, in *News from the Land of Freedom*, 281; "Foreign Languages in the Elementary School," *School and Society* 6, no. 151 (1917): 583–584.

20. Carolyn R. Toth, *German-English Bilingual Schools in America: The Cincinnati Tradition in Historical Context* (New York: Peter Lang, 1990), 83; "Foreign Languages in the Elementary School," 584; Steven L. Schlossman, "Is There an American Tradition of Bilingual Education?: German in the Public Elementary Schools, 1840–1919," *American Journal of Education* 91, no. 2 (1983): 176–178; Milwaukee Public Schools, *Fifty-Seventh Annual Report of the Board of School Directors of the City of Milwaukee for the Year Ending June 30, 1916* (Milwaukee: Radtke Bros. and Kortsch Co., n.d.), 61–63; Milwaukee Public Schools, *Fifty-Eighth Annual Report of the Board of School Directors of the City of Milwaukee for the Year Ending June 30, 1917* (Milwaukee: Radtke Bros. and Kortsch Co., n.d.), 19–20; Sommerfeld, "Elimination of German from the Cleveland School," 276–278.

21. Cincinnati Public Schools, *Eighty-Eighth Annual Report for the School Year Ending August 31, 1917* (Cincinnati: n.p., 1918), 109, 344; Toth,

German-English Bilingual Schools in America, 87–90; "Foreign Languages in the Elementary School," 583–584.

22. Frances H. Ellis, "German Instruction in the Public Schools of Indianapolis, 1869–1919: III," *Indiana Magazine of History* 50, no. 4 (1954): 368–375; Ramsey, "The War against German-American Culture," 298; Indianapolis Public Schools, *Annual Report: Business Director, Superintendent of Schools and Librarian* (n.p., 1916), 130–132.

23. Ellis, "German Instruction in the Public Schools of Indianapolis," 373; David A. Gerber, "Language Maintenance, Ethnic Group Formation, and Public Schools: Changing Patterns of German Concern, Buffalo, 1837–1874," *Journal of American Ethnic History* 4, no. 1 (1984): 49; Luebke, *Germans in the New World*, 32–47; I. N. Edwards, "The Legal Status of Foreign Languages in the Schools," *The Elementary School Journal* 24, no. 4 (1923): 270–273; Henry J. Fletcher, ed., "Recent Legislation Forbidding Teaching of Foreign Languages in Public Schools," *Minnesota Law Review* 4, no. 6 (1920): 449–551.

24. Edwards, "The Legal Status of Foreign Languages in the Schools," 272; Ramsey, "The War against German-American Culture," 299–301; *Journal of the House of Representatives of the State of Indiana during the Seventy-First Session of the General Assembly Commencing Thursday, January 9, 1919* (Indianapolis: Wm. B. Burford, 1919), 519–520; "House Passes M'Cray's Bill in 15 Minutes," *Indianapolis Star*, February 26, 1919, 1, 5.

25. Luebke, *Germans in the New World*, 41–46; Kenneth B. O'Brien, Jr., "Education, Americanization and the Supreme Court: The 1920s," *American Quarterly* 13, no. 2 (1961): 161–166; Ramsey, "The War against German-American Culture," 301–302; Edwards, "The Legal Status of Foreign Languages in the Schools," 270–278; "Recent Legislation Forbidding Teaching of Foreign Languages in Public Schools," 450; *Meyer v. Nebraska*, 262 U.S. 390 (1923).

26. Edwards, "The Legal Status of Foreign Languages in the Schools," 272; Carlos Kevin Blanton, *The Strange Career of Bilingual Education in Texas, 1836–1981* (College Station: Texas A&M University Press, 2004), 62–77; Guadalupe San Miguel, Jr., *"Let All of Them Take Heed": Mexican Americans and the Campaign for Educational Equality in Texas, 1910–1981* (Austin: University of Texas Press, 1987), 32–37; W. F. Doughty and R. B. Binnion, *School Laws of Texas, 1913* (Austin: Von [Boe]ckmann-Jones Company, Printers, n.d.), 82–83.

27. U.S. Bureau of Education, *A Survey of Education in Hawaii* (Washington, DC: Government Printing Office, 1920), 130–131; Eileen H. Tamura, "The English-Only Effort, the Anti-Japanese Campaign, and Language Acquisition in the Education of Japanese Americans in Hawaii, 1915–40," *History of Education Quarterly* 33, no. 1 (1993): 37–45; Noriko Asato, *Teaching Mikadoism: The Attack on Japanese Language Schools in Hawaii, California, and Washington, 1919–1927* (Honolulu: University of Hawai'i Press, 2006), 7–14, 21–79, 99–100.

28. Robert F. Roeming, "Bilingualism and the National Interest," *The Modern Language Journal* 55, no. 2 (1971): 75–78; David P. Benseler, "In Memoriam:

Robert F. Roeming, 1912–2004," *The Modern Language Journal* 89, no. 2 (2005):159–160.

29. "Benz Explains Vote," *Indianapolis Star*, February 26, 1919, 1, 5; "Quick Action Demanded," *Indianapolis Star*, February 26, 1919, 5; Ramsey, "The War against German-American Culture," 299; A. J. P. Taylor, *The Origins of the Second World War* (1961; reprint, with a new introduction, New York: Touchstone, 1996), 18–39.

8 Rebuilding the Boardinghouse: The Interwar Years

1. Theodore Andersson, "Bilingual Education: The American Experience," *The Modern Language Journal* 55, no. 7 (1971): 428; Theodore Andersson, "Bilingual Education: The American Experience" (paper presented at the Conference on Bilingual Education, Ontario Institute for Studies in Education, Toronto, March 13, 1971), box 1, file 16, Theodore Anderson Bilingual Education papers, Nettie Lee Benson Latin American Collection, University of Texas, Austin (hereafter cited as NLB Collection); Theodore Andersson, "Testimony Presented on H.R. 9840 and H.R. 10224 to Authorize Bilingual Education Programs in Elementary and Secondary Schools before the House General Subcommittee on Education and Labor," June 29, 1967, box 1, file 14, Theodore Anderson Bilingual Education papers, NLB Collection; Carlos Kevin Blanton, *The Strange Career of Bilingual Education in Texas, 1836–1981* (College Station: Texas A&M University Press, 2004), 121–136.

2. Herbert Hoover, "True Liberalism Seeks All Legitimate Freedom," in *Our Nation's Archive: The History of the United States in Documents*, ed. Erik Bruun and Jay Crosby (New York: Tess Press, 1999), 678–681; Wall Street Journal, "Demoralization Was Unprecedented," in *Our Nation's Archive*, 681; Ludwig Dilger, November 7, 1925, in *News from the Land of Freedom: German Immigrants Write Home*, ed. Walter D. Kamphoefner, Wolfgang Helbich, and Ulrike Sommer, trans. Susan Carter Vogel (Ithaca: Cornell University Press, 1991), 508; Ludwig Dilger, February 12, 1926, in *News from the Land of Freedom*, 508–509; David Tyack, Robert Lowe, and Elisabeth Hansot, *Public Schools in Hard Times: The Great Depression and Recent Years* (Cambridge, MA: Harvard University Press, 1984), 6–7.

3. Studs Terkel, "Recalling the Depression," in *Our Nation's Archive*, 684–686; Franklin D. Roosevelt, "The Only Thing We Have to Fear Is Fear Itself," in *Our Nation's Archive*, 687–688; Ludwig Dilger, December 11, 1931, in *News from the Land of Freedom*, 515; Ludwig Dilger, October 10, 1932, in *News from the Land of Freedom*, 516; Tyack, Lowe, and Hansot, *Public Schools in Hard Times*, 6.

4. Richard Hofstadter, ed., *The Progressive Movement, 1900–1915* (New York: Touchstone, 1986); Richard Hofstadter, *The Age of Reform: From Bryan to F. D. R.* (New York: Vintage Books, 1955), 3–22; Henry F. May, *The End of American Innocence: A Study of the First Years of Our Own Time, 1912–1917* (New York: Alfred A. Knopf, 1959); Morton Keller,

Regulating a New Society: Public Policy and Social Change in America, 1900–1933 (Cambridge, MA: Harvard University Press, 1994), 5–6; Randolph Bourne, "Trans-National America," in *The American Intellectual Tradition: A Sourcebook*, ed. David A. Hollinger and Charles Capper, vol. 2, 4th ed. (New York: Oxford University Press, 2001), 171–181; David A. Hollinger, *Postethnic America: Beyond Multiculturalism* (New York: Basic Books, 1995), 84–86, 92–94; "The National Recovery Act," in *Our Nation's Archive*, 691–695; "The Social Security Act of 1935," in *Our Nation's Archive*, 716–722; Tyack, Lowe, and Hansot, *Public Schools in Hard Times*, 92–138; Gary Gerstle, *American Crucible: Race and Nation in the Twentieth Century* (Princeton: Princeton University Press, 2001), 131–155.

5. Keller, *Regulating a New Society*, 43–66; Tyack, Lowe, and Hansot, *Public Schools in Hard Times*, 27–41; Jeffrey Mirel, *The Rise and Fall of an Urban School System: Detroit, 1907–81* (Ann Arbor: University of Michigan Press, 1993), 55–66, 89–137.

6. Herbert M. Kliebard, *The Struggle for the American Curriculum, 1893–1958*, 3rd ed. (New York: RoutledgeFalmer, 2004), 7, 36–44, 51–75, 92, 130–150, 158–161; Lawrence A. Cremin, *The Transformation of the School: Progressivism in American Education, 1876–1957* (New York: Vintage Books, 1964), 100–105, 111–126, 305–306; Larry Cuban, *How Teachers Taught: Constancy and Change in American Classrooms, 1880–1990*, 2nd ed. (New York: Teachers College Press, 1993), 45–52, 112–114.

7. Kliebard, *The Struggle for the American Curriculum*, 76–104, 106; Herbert M. Kliebard, *Schooled to Work: Vocationalism and the American Curriculum, 1876–1946* (New York: Teachers College Press, 1999), 153–162; David Nasaw, *Schooled to Order: A Social History of Public Schooling in the United States* (Oxford: Oxford University Press, 1981), 126–145; Tyack, Lowe, and Hansot, *Public Schools in Hard Times*, 92–138; Mirel, *The Rise and Fall of an Urban School System*, 89–137 Paul W. Horn, *Survey of the City Schools of El Paso Texas* (El Paso: Department of Printing of the City Schools, El Paso, Texas, 1922), 33–35; Texas Educational Survey Commission, *Texas Educational Survey Report: Courses of Study and Instruction* (Austin: Texas Educational Survey Commission, 1924).

8. Jane Addams, "Educational Methods," in *On Education*, ed. Ellen Condliffe Lagemann (New Brunswick: Transaction Publishers, 2002), 98–100, 102; John Dewey, *Democracy and Education: An Introduction to the Philosophy of Education* (1916; reprint, New York: Free Press, 1997), 87; George S. Counts, *Dare the School Build a New Social Order?* (1932; reprint, with a new preface by Wayne J. Urban, Carbondale: Southern Illinois University Press, 1978), 38, 44–52; Kliebard, *The Struggle for the American Curriculum*, 151–174; Social Frontier Editors, "Orientation," in *The School in the United States: A Documentary History*, ed. James W. Fraser (Boston: McGraw-Hill, 2001), 218–219; John Dewey, "Can Education Share in Social Reconstruction?," in *The School in the United States*, 219–221.

9. Counts, *Dare the School Build a New Social Order?*, 4–5, 22–23; B. Edward McClellan, *Moral Education in America: Schools and the Shaping of*

Character from the Colonial Times to the Present (New York: Teachers College Press, 1999), 46–69; Paul J. Ramsey, "'Let Virtue Be Thy Guide, and Truth Thy Beacon-Light": Moral and Civic Transformation in Indianapolis's Public Schools," in *Civic and Moral Learning in America*, ed. Donald Warren and John J. Patrick (New York: Palgrave Macmillan, 2006), 135–151; Addams, "Educational Methods," in *On Education*, 102; Indianapolis Public Schools, *Annual Report of the Secretary, Business Director, Superintendent of Schools and the Librarian* (n.p., [1909?]), 35–39; Sinclair Lewis, *Babbitt* (1922; reprint, Mineola, NY: Dover Publications, Inc., 2003), 59–61, 148, 156; Tyack, Lowe, and Hansot, *Public Schools in Hard Times*, 18–27.

10. Cuban, *How Teachers Taught*, 272–282; Dolly Holliday Clark, "Memoir of a Country Schoolteacher: Dolly Holliday Meets the Ethnic West, 1919–1920," ed. Paula M. Nelson, *North Dakota History* 59, no. 1 (1992): 30–45; Walter D. Kamphoefner, "German American Bilingualism: *cui malo?* Mother Tongue and Socioeconomic Status among the Second Generation in 1940," *International Migration Review* 28, no. 4 (1994): 846–864.

11. Else Hoffmann, "The German-Language School in Baltimore," *The German-American Review* 5, no. 2 (1938): 34–35, 55.

12. I. N. Edwards, "The Legal Status of Foreign Languages in the Schools," *The Elementary School Journal* 24, no. 4 (1923): 272; "Esther Vaagen," *North Dakota History* 44, no. 4 (1977): 14, 16; L. D. Coffman, *Illinois School Survey: A Coöperative Investigation of School Conditions and School Efficiency, Initiated and Conducted by the Teachers of Illinois in the Interest of All the Children and All the People* (Illinois State Teachers Association, 1917), 28–30 82–83, 88–91.

13. *Meyer v. Nebraska*, 262 U.S. 390 (1923); Kenneth B. O'Brien, Jr., "Education, Americanization and the Supreme Court: The 1920's," *American Quarterly* 13, no. 2 (1961): 163–166.

14. *Pierce v. Society of Sisters*, 268 U.S. 510 (1925); O'Brien, "Education, Americanization and the Supreme Court," 166–168.

15. John Higham, *Strangers in the Land: Patterns of American Nativism, 1860–1925* (New Brunswick, NJ: Rutgers University Press, 1955), 202–203; Gerstle, *American Crucible*, 3–11, 81–127; Prescott F. Hall, "Immigration and the World War," *Annals of the American Academy of Political and Social Science* 93 (January 1921): 190–193; Lewis, *Babbitt*, 110.

16. U.S. Congress, "The Immigration Act of 1924," in *Our Nation's Archive*, 657–660; Higham, *Strangers in the Land*, 300–330; Stuart Ewen and Elizabeth Ewen, *Typecasting: On the Arts and Sciences of Human Inequality* (New York: Seven Stories Press, 2006), 308–310.

17. Leonard Covello, "Language as a Factor in Integration and Assimilation: The Role of the Language Teacher in a School-Community Program," *The Modern Language Journal* 23, no. 5 (1939): 323–324; Joshua A. Fishman, *Hungarian Language Maintenance in the United States* (Bloomington: Indiana University, 1966), 8–10.

18. Fishman, *Hungarian Language Maintenance in the United States*, 11–12; "Varia: French in the Public Elementary Schools in Louisiana," *The French*

Review 13, no. 4 (1940): 344–346; Covello, "Language as a Factor in Integration and Assimilation," 326–331; Mario E. Cosenza, Italian Teachers Association, "Eighth Annual Report, School Year 1928–1929," in *The Italian Community and Its Language in the United States: The Annual Reports of the Italian Teachers Association,* ed. Francesco Cordasco (Totowa, NJ: Rowman and Littlefield, 1975), 79–88; Leonard Covello, Italian Teachers Association, "Eighteenth Annual Report, School Year 1938–1939," in *The Italian Community and Its Language in the United States,* 359, 377–383; Cuban, *How Teachers Taught,* 63–64; Selma Cantor Berrol, *Growing Up American: Immigrant Children in America, Then and Now* (New York: Twayne Publishers, 1995), 91–95.

19. Ronald Takaki, *A Different Mirror: A History of Multicultural America* (Boston: Little, Brown and Company, 1993), 340–347; James D. Anderson, *The Education of Blacks in the South, 1860–1935* (Chapel Hill: University of North Carolina Press, 1988), 260–261; Toni Morrison, *The Bluest Eye* (New York: Vintage International, 2007), 111, 116; Steven A. Reich, "The Great Migration and the Literary Imagination," *The Journal of the Historical Society* 9, no. 1 (2009): 88–89.

20. Matthias Dorgathen, June 20, [1883?], in *News from the Land of Freedom,* 454; Noel Ignatiev, *How the Irish Became White* (New York: Routledge Classics, 2009), 170–203; Russell A. Kazal, *Becoming Old Stock: The Paradox of German-American Identity* (Princeton: Princeton University Press, 2004), 6; Hasia R. Diner, "The World of Whiteness," *Historically Speaking* 9, no. 1 (2007): 20–22; Gerstle, *American Crucible,* 162–176.

21. Nicholas V. Montalto, "The Intercultural Education Movement, 1924–41: The Growth of Tolerance as a Form of Intolerance," in *Education and the European Immigrant, 1840–1940,* ed. Bernard J. Weiss (Urbana: University of Illinois Press, 1982), 142–160; Covello, "Language as a Factor in Integration and Assimilation," 323–325; Noel Epstein, *Language, Ethnicity, and the Schools: Policy Alternatives for Bilingual-Bicultural Education* (Washington, DC: Institute for Educational Leadership, 1977), 19–20.

22. John M. Nieto-Phillips, *The Language of Blood: The Making of Spanish-American Identity in New Mexico, 1880s-1930s* (Albuquerque: University of New Mexico Press, 2004), 1–11; George I. Sanchez, *Forgotten People: A Study of New Mexicans* (Albuquerque: University of New Mexico Press, 1940), 12–14, 27; Frances Esquibel Tywoniak and Mario T. García, *Migrant Daughter: Coming of Age as a Mexican American Woman* (Berkley: University of California Press, 2000), 1–6.

23. Nieto-Phillips, *Language of Blood,* 197–205; Isabel Lancaster Eckles, *Thirty-Third and Thirty-Fourth Annual Reports of the State Superintendent of Public Instruction to the Governor of New Mexico, 1923–1924* (n.p., n.d.), 7–11, 78–79; Isabel Lancaster Eckles, *Thirty-Fifth and Thirty-Sixth Annual Reports of the State Superintendent of Public Instruction to the Governor of New Mexico, 1925–1926* (n.p., n.d.), 7–11, 74–75.

24. Nieto-Phillips, *Language of Blood,* 200–203; George Isidore Sanchez, "The Education of Bilinguals in a State School System" (PhD diss., University of California, 1934), 42, 80–82, 98, 100–103; Annie Reynolds, *The Education*

of *Spanish-Speaking Children in Five Southwestern States* (Washington, DC: Government Printing Office, 1933), 26–29; Esquibel Tywoniak and García, *Migrant Daughter*, 1–2, 23–24.

25. U.S. Congress, "The Immigration Act of 1924," in *Our Nation's Archive*, 658–659; Higham, *Strangers in the Land*, 324; Takaki, *A Different Mirror*, 311–339; Manuel G. Gonzales, *Mexicanos: A History of Mexicans in the United States* (Bloomington: Indiana University Press, 2000), 113–138; Julian Nava, *My Mexican-American Journey* (Houston: Arte Público Press, 2002), 3; Esquibel Tywoniak and García, *Migrant Daughter*, 9–12.

26. Takaki, *A Different Mirror*, 311–339; Gonzales, *Mexicanos*, 113–138; Nava, *My Mexican-American Journey*, 2–4, 79.

27. Sandra Cisneros, *The House on Mango Street* (New York: Vintage Contemporaries, 1991), 76–78; Herschel T. Manuel, *The Education of Mexican and Spanish-Speaking Children in Texas* (Austin: University of Texas, 1930), 18; Esquibel Tywoniak and García, *Migrant Daughter*, 21–23, 42–44, 51–54; Guadalupe San Miguel, Jr., *"Let All of Them Take Heed": Mexican Americans and the Campaign for Educational Equality in Texas, 1910–1981* (Austin: University of Texas, 1987), 70–71; Gloria López-Stafford, *A Place in El Paso: A Mexican-American Childhood* (Albuquerque: University of New Mexico Press, 1996), 6–7, 43–46.

28. Ruben Donato, "Hispano Education and the Implications of Autonomy: Four School Systems in Southern Colorado, 1920–1963," *Harvard Educational Review* 69, no. 2 (1999): 117–149; Esquibel Tywoniak and García, *Migrant Daughter*, 56–57; Reynolds, *The Education of Spanish-Speaking Children*, 9–13; Sanchez, "The Education of Bilinguals," 33–38; Manuel, *The Education of Mexican and Spanish-Speaking Children*, 1–37, 56–90; George I. Sanchez, *Concerning Segregation of Spanish-Speaking Children in the Public Schools* (Austin: University of Texas, 1951); Rubén Donato, *The Other Struggle for Equal Schools: Mexican Americans during the Civil Rights Era* (Albany: State University of New York Press, 1997), 12–17; Matthew D. Davis, *Exposing a Culture of Neglect: Herschel T. Manuel and Mexican American Schooling* (Greenwich, CT: Information Age Publishing, 2005), 63–75.

29. Donato, *The Other Struggle for Equal Schools*, 13; Blanton, *Strange Career*, 78–91; Nava, *My Mexican-American Journey*, 9–10; Reynolds, *The Education of Spanish-Speaking Children*, 50–51; Esquibel Tywoniak and García, *Migrant Daughter*, 41–42.

30. Reynolds, *The Education of Spanish-Speaking Children*, 58–59; Manuel, *The Education of Mexican and Spanish-Speaking Children*, 124–126; Heinz Kloss, *The American Bilingual Tradition* (1977; reprint, with a new introduction by Reynaldo F. Macías and Terrence G. Wiley, McHenry, IL: Center for Applied Linguistics and Delta Systems Co., Inc., 1998), 119; Blanton, *Strange Career*, 76–79, 89–90; San Miguel, *Let All of Them Take Heed*, 37, 66–67.

31. Jon Reyhner and Jeanne Eder, *American Indian Education: A History* (Norman: University of Oklahoma Press, 2004), 205–215; David Wallace Adams, *Education for Extinction: American Indians and the Boarding*

School Experience, 1875–1928 (Lawrence: University Press of Kansas, 1995), 330–333; Institute for Government Research, *The Problem of Indian Administration* (Baltimore: Johns Hopkins Press, 1928), 346, 372–373; "Indian Citizenship Act: June 2, 1924," in *Documents of United States Indian Policy,* ed. Francis Paul Prucha, 3rd ed. (Lincoln: University of Nebraska Press, 2000), 218.

32. Reyhner and Eder, *American Indian Education,* 205–231; K. Tsianina Lomawaima and Teresa L. McCarty, *"To Remain an Indian": Lessons in Democracy from a Century of Native American Education* (New York: Teachers College Press, 2006), 6–7, 91–113; "Wheeler-Howard Act (Indian Reorganization Act): June 18, 1934," in *Documents of United States Indian Policy,* 223–225; "Indian Commissioner Collier on the Wheeler-Howard Act," in *Documents of United States Indian Policy,* 225–229; "Oklahoma Indian Welfare Act: June 26, 1936," in *Documents of United States Indian Policy,* 230–231; "Indian Arts and Crafts Board: August 27, 1935," in *Documents of United States Indian Policy,* 229–230; Cuban, *How Teachers Taught,* 49.

33. Reyhner and Eder, *American Indian Education,* 205–231; Lomawaima and McCarty, *To Remain an Indian,* 91–149; Ann Nolan Clark, *In My Mother's House* (1941; reprint, New York: Viking Press, 1966); Virginia Woolf, *Jacob's Room* (1922; reprint, with an introduction by Danell Jones, New York: Barnes & Noble, 2006), 37; Bernard Spolsky, *American Indian Bilingual Education* (University of New Mexico, 1974), 1–38.

Epilogue: The Federal Landlord

1. Theodore Andersson and Mildred Boyer, *Bilingual Schooling in the United States: History, Rationale, Implications, and Planning,* vol. 1 (Austin: Southwest Educational Development Laboratory, 1970); Joshua A. Fishman, *Language Loyalty in the United States: The Maintenance and Perpetuation of Non-English Mother Tongues by American Ethnic and Religious Groups* (London: Mouton & Co., 1966); Guadalupe San Miguel, Jr., *"Let All of Them Take Heed": Mexican Americans and the Campaign for Educational Equality in Texas, 1910–1981* (Austin: University of Texas, 1987), 172–186; 195–201; Guadalupe San Miguel, Jr., *Contested Policy: The Rise and Fall of Federal Bilingual Education in the United States, 1960–2001* (Denton, TX: University of North Texas Press, 2004), 5–53; James Crawford, *Bilingual Education: History, Politics, Theory, and Practice,* 3rd ed. (Los Angeles: Bilingual Educational Services, 1995), 36–53; Rubén Donato, *The Other Struggle for Equal Schools: Mexican Americans during the Civil Rights Era* (Albany: State University of New York Press, 1997), 103–118; Diego Castellanos, *The Best of Two Worlds: Bilingual-Bicultural Education in the U.S.,* 2nd printing (Trenton: New Jersey State Department of Education, 1985), 47–86.

2. San Miguel, *Contested Policy,* 5–53; Crawford, *Bilingual Education,* 36–53; Donato, *The Other Struggle for Equal Schools,* 103–118; Castellanos, *The*

Best of Two Worlds, 47–86; Heinz Kloss, *The American Bilingual Tradition* (1977; reprint, with a new introduction by Reynaldo F. Macías and Terrence G. Wiley, McHenry, IL: Center for Applied Linguistics and Delta Systems Co., Inc., 1998), 40–45; Carlos Kevin Blanton, *The Strange Career of Bilingual Education in Texas, 1836–1981* (College Station: Texas A&M University Press, 2004), 141–152; Timothy A. Hacsi, *Children as Pawns: The Politics of Educational Reform* (Cambridge, MA: Harvard University Press, 2002), 69–74.

3. Kloss, *The American Bilingual Tradition*, 39–40; Crawford, *Bilingual Education*, 36–53; Castellanos, *The Best of Two Worlds*, 47–86; "National Defense Education Act, 1958," in *The School in the United States: A Documentary History*, ed. James W. Fraser (Boston: McGraw-Hill, 2001), 224–225; H. G. Rickover, "Education for All Children: What We Can Learn from England , 1962," in *The School in the United States*, 229–232; H. G. Rickover, *American Education—A National Failure: The Problem of Our Schools and What We Can Learn from England* (New York: E. P. Dutton and Co., Inc., 1963); James B. Conant, *The American High School Today: A First Report to Interested Citizens* (New York: McGraw-Hill Book Company, Inc., 1960).

4. Kloss, *The American Bilingual Tradition*, 37–45; San Miguel, *Contested Policy*, 5–53; Donato, *The Other Struggle for Equal Schools*, 50–85; Richard Neumann, *Sixties Legacy: A History of the Public Alternative Schools Movement, 1967–2001* (New York: Peter Lang, 2003), 9–72; Paul Goodman, *Compulsory Mis-education* (New York: Horizon Press, 1964); Paul Cowan, *The Tribes of America: Journalistic Discoveries of Our People and Their Cultures* (New York: New Press, 2008), 90; Supreme Court of the United States, "Brown v. Board of Education of Topeka, Kansas, 1954," in *The School in the United States*, 266–269; "The Elementary and Secondary Education Act and the Great Society, 1965," in *The School in the United States*, 295–298.

5. George I. Sanchez to Lee Wilborn, February 16, 1960, box 26, file 2, George I. Sanchez papers, NLB Collection. For the intense work of these scholars, see Joshua A. Fishman to George I. Sanchez, January 6, 1959, box 15, file 18, George I. Sanchez papers, NLB Collection; Joshua A. Fishman to George I. Sanchez, May 15, 1961, box 15, file 18, George I. Sanchez papers, NLB Collection; Joshua A. Fishman to George I. Sanchez, June 5, 1961, box 15, file 18, George I. Sanchez papers, NLB Collection; George I. Sanchez to Joshua A. Fishman, June 20, 1961, box 15, file 18, George I. Sanchez papers, NLB Collection; Joshua A. Fishman to George I. Sanchez, October 19, 1961, box 15, file 18, George I. Sanchez papers, NLB Collection; George I. Sanchez to Joshua Fishman and Theodore Andersson, June 17, 1963, box 15, file 18, George I. Sanchez papers, NLB Collection.

6. Julian Nava, *My Mexican-American Journey* (Houston: Arte Público Press, 2002), 76–77; Ralph Yarborough, "Introducing the Bilingual Education Act," in *Language Loyalties: A Source Book on the Official English Controversy*, ed. James Crawford (Chicago: University of Chicago Press, 1992), 324; Theodore Andersson, "Testimony Presented on H.R. 9840 and

H.R. 10224 to Authorize Bilingual Education Programs in Elementary and Secondary Schools before the House General Subcommittee on Education and Labor," June 29, 1967, box 1, file 14, Theodore Anderson Bilingual Education papers, NLB Collection; Blanton, *Strange Career*, 124–140; San Miguel, *Contested Policy*, 1–25; Pablo Medina, *Exiled Memories: A Cuban Childhood* (New York: Persea Books, 2002), 108–109; Francesco Cordasco, ed., *Dictionary of American Immigrant History* (Metuchen, NJ: Scarecrow Press, 1990), 151–159.

7. San Miguel, *Contested Policy*, 1–25; Hacsi, *Children as Pawns*, 75–102; Jonathan Kozol, *The Shame of the Nation: The Restoration of Apartheid Schooling in America* (New York: Crown Publishers, 2005), 1–12.

Bibliography

Unpublished Sources

George I. Sanchez Papers. Nettie Lee Benson Latin American Collection. University of Texas, Austin, TX.

Harris, W. T. "German Instruction in American Schools and the National Idiosyncrasies of the Anglos-Saxons and the Germans." Paper presented at the National German-American Teachers' Association, Cleveland, OH, July 16, 1890. University of Illinois Library, Urbana, IL.

Peaslee, John B. "Instruction in German and Its Helpful Influence on Common-School Education as Experienced in the Public Schools of Cincinnati." Paper presented at the National German-American Teachers' Association, Chicago, IL, July 19, 1889. University of Illinois Library, Urbana, IL.

Theodore Andersson Bilingual Education Papers. Nettie Lee Benson Latin American Collection. University of Texas, Austin, TX.

School Reports, Laws, and Surveys: Local, State, and Territorial

Belleville Public Schools. *First Annual Report for the School Year Ending June 26, 1874, and Various Supplementary Documents Exhibiting the Condition of the Schools*. Stern des Westens, 1874.

Biennial Report of the Superintendent of Public Instruction of the Territory of Arizona. N.p., 1890.

Bobbitt, J. F. *The San Antonio Public School System: A Survey*. San Antonio: San Antonio School Board, 1915.

By-Laws of the School Officers and Trustees of the Nineteenth Ward, and Rules for the Government of Schools. New York: J. Youdale, 1863.

California Council of Education. *First Report of the Committee on the Course of Study for Elementary Schools*. Oakland: N.p., 1896.

Cincinnati Public Schools. *Eighty-Eighth Annual Report for the School Year Ending August 31, 1917*. Cincinnati: n.p., 1918.

Cleveland Public Schools. *Thirty-Seventh Annual Report of the Board of Education for the School Year Ending Aug. 31, 1873*. Cleveland: Leader Printing Company, 1874.

Cleveland Public Schools. *Thirty-Eighth Annual Report of the Board of Education for the School Year Ending August 31, 1874.* Cleveland: Robison, Savage and Company, 1875.

Coffman, L. D. *Illinois School Survey: A Coöperative Investigation of School Conditions and School Efficiency, Initiated and Conducted by the Teachers of Illinois in the Interest of All the Children and All the People.* Illinois State Teachers Association, 1917.

Common Schools of Cincinnati. *Thirty-Eighth Annual Report for the School Year Ending June 30, 1867.* Cincinnati: Gazette Steam Book and Job Printing Establishment, 1867.

———. *Forty-Second Annual Report for the School Year Ending June 30, 1871.* Cincinnati: Wilstach, Baldwin and Co., 1872.

———. *Fifty-Seventh Annual Report for the School Year Ending August 31st, 1886.* Cincinnati: Ohio Valley Publishing and Manufacturing Company, 1887.

Common Schools of Indianapolis. *Annual Report of the Public Schools of the City of Indianapolis, for the School Year Ending Sept. 1, 1866.* Indianapolis: Douglass and Conner, 1867.

Department of Education, the City of New York. *Fourth Annual Report of the City Superintendent of Schools for the Year Ending July 31, 1902.* New York: N.p., n.d. Doughty, W. F., and R. B. Binnion. *School Laws of Texas, 1913.* Austin: Von [Boe]ckmann-Jones Company, N.d.

Eckles, Isabel Lancaster. *Thirty-Third and Thirty-Fourth Annual Reports of the State Superintendent of Public Instruction to the Governor of New Mexico, 1923–1924.* N.p., n.d.

———. *Thirty-Fifth and Thirty-Sixth Annual Reports of the State Superintendent of Public Instruction to the Governor of New Mexico, 1925–1926.* N.p., n.d.

Eighth Biennial Report of the Superintendent of Public Instruction: State of South Dakota, 1905–1906. Huron, SD: John Longstaff, n.d.

Eleventh Biennial Report of the Superintendent of Public Instruction of the State of Illinois for the Two Years Ending Sept. 30, 1876. Springfield: D. W. Lusk, State Printers and Binders, 1877.

Fifteenth Annual Report of the Belleville Public Schools for the School Year Ending June 22, 1888, and Various Supplementary Documents, Exhibiting the Conditions of the Schools. Geo. Semmelroth and Co., n.d.

Fifteenth Biennial Report of the State Superintendent of Public Instruction to the Governor of the State of Nebraska, Dec. 1, 1898. Lincoln: Jacob North and Co., 1898.

Forty-Eighth Annual Report of the Board of Commissioners of Public Schools, to the Mayor and City Council of Baltimore, for the Year Ending October 31st, 1876. Baltimore: King Brothers, 1877.

Forty-Fifth Annual Report of the Board of Commissioners of Public Schools, to the Mayor and City Council of Baltimore, for the Year Ending October 31st, 1873. Baltimore: King Brothers, 1874.

Forty-First Annual Report of the Superintendent of Public Instruction of the State of Michigan, with Accompanying Documents, for the Year 1877. Lansing: W. S. George and Co., 1878.

Forty-First Report of the Belleville Public Schools, District No. 118, for the School Years Ending June 22, 1917 and June 21, 1918. N.p., 1918.

Forty-Seventh Annual Report of the Board of Commissioners of Public Schools, to the Mayor and City Council of Baltimore, for the Year Ending October 31, 1875. Baltimore: King Brothers, 1875.

Fourteenth Annual Report of the Superintendent of Public Instruction of the State of New York. Albany: Van Benthuysen and Sons' Steam Printing House, 1868.

Fourth Biennial Report of the Superintendent of Public Instruction to the Governor of North Dakota, for the Two Years Ending June 30, 1896. Jamestown, ND: Alert, State Printers and Binders, 1896.

Henderson, Howard A. M. *The Kentucky School-Lawyer: Or A Commentary on the Kentucky School Laws and Regulations of the State Board of Education.* Frankfort, KY: Major, Johnston and Barrett, 1877.

Hines, L. N., ed. *Proceedings and Papers of the Indiana State Teachers' Association, October 31, and November 1, 2, 3, 1917.* N.p., n.d.

Horn, Paul W. *Survey of the City Schools of El Paso Texas.* El Paso: Department of Printing of the City Schools, El Paso, Texas, 1922.

Indianapolis Public Schools. *Annual Report of the Secretary, Business Director, Superintendent of Schools and the Librarian.* N.p., 1902.

———. *Annual Report of the Secretary, Business Director, Superintendent of Schools and the Librarian.* N.p., [1909].

———. *Annual Report: Business Director, Superintendent of Schools and Librarian.* N.p., 1916.

Journal of the House of Representatives of the State of Indiana during the Seventy-First Session of the General Assembly Commencing Thursday, January 9, 1919. Indianapolis: Wm. B. Burford, 1919.

Laws of the State of Indiana Passed at the Seventy-First Regular Session of the General Assembly. Indianapolis: Wm. B. Burford, 1919.

Louisiana State Department of Education. *Public School Statistics, Session of 1914–'15: Part II of the Biennial Report of 1913–'14 and 1914–'15.* Baton Rouge: Ramires-Jones Printing Company, 1915.

A Manual of Discipline and Instruction for the Use of the Teachers of the Primary and Grammar Schools under the Charge of the Department of Public Instruction of the City of New York. New York: Evening Post Steam Presses, 1873.

Miller, Herbert Adolphus. *Cleveland Educational Survey: The School and the Immigrant.* Cleveland: Survey Committee of the Cleveland Foundation, 1916.

Milwaukee Public Schools. *Annual Report of the School Board of the City of Milwaukee, for the Year Ending August 31, 1876.* Milwaukee: N.p., 1876.

———. *Fifty-Seventh Annual Report of the Board of School Directors of the City of Milwaukee for the Year Ending June 30, 1916.* Milwaukee: Radtke Bros. and Kortsh Co., n.d.

———. *Fifty-Eighth Annual Report of the Board of School Directors of the City of Milwaukee for the Year Ending June 30, 1917.* Milwaukee: Radtke Bros. and Kortsch Co., n.d.

Minority Report of the Committee on Text-Books, on Text-Books in History in the High Schools. Boston: Rockwell and Churchill, City Printers, 1890.

Nineteenth and Twentieth Annual Reports of the Territorial Superintendent of Public Instruction to the Governor of New Mexico for the Years 1909–1910. Santa Fe: New Mexican Printing Company, 1911.

Pennsylvania Common Schools. *Report of the Superintendent of Common Schools of Pennsylvania, for the Year Ending June 2, 1857.* Harrisburg: A. Boyd Hamilton, 1858.

————. *Report of the Superintendent of Common Schools of Pennsylvania, for the Year Ending June 3, 1861.* Harrisburg: A. Boyd Hamilton, 1862.

————. *Report of the Superintendent of Common Schools of Pennsylvania, for the Year Ending June 2, 1862.* Harrisburg: Singerly and Myers, 1863.

————. *Report of the Superintendent of Common Schools of Pennsylvania, for the Year Ending June 4, 1863.* Harrisburg: Singerly and Myers, 1864.

————. *Report of the Superintendent of Common Schools of the Commonwealth of Pennsylvania, for the Year Ending June 3, 1867.* Harrisburg: Singerly and Myers, 1868.

Public Schools of Cincinnati. *Eighty-Second Annual Report of the Public Schools of Cincinnati for the School Year Ending August 31, 1911.* N.p., n.d.

Report of the Superintendent of Public Instruction of the Commonwealth of Kentucky, with Summaries of Statistics from the School-Year Ending June 30th, 1881, to the School-Year, 1886. Frankfort, KY: John D. Woods, Public Printer and Binder, 1886.

Report of the Superintendent [of] Public Instruction of the State of Kentucky for Four Scholastic Years Ended June 30, 1891. Frankfort, KY: E. Polk Johnson, Public Printer and Binder, 1892.

Rules of the School Committee, and Regulations of the Public Schools of the City of Boston. Boston: Geo. C. Rand and Avery, 1860.

Second Report of the Board of Education of Louisville, Kentucky from July 1, 1912 to June 30, 1913. Louisville: Gross Parsons and Hambleton, n.d.

Seventeenth Annual Report of the Board of Directors of the St. Louis Public Schools, for the Year Ending August 1, 1871. St. Louis: Plate, Olshausen and Co., 1872.

Seventeenth and Eighteenth Annual Reports of the Territorial Superintendent of Public Instruction to the Governor of New Mexico for the Years 1907–1908. Santa Fe: New Mexican Printing Company, 1909.

Sixteenth Annual Report of the Superintendent of Public Instruction of the State of New York. Albany: Argus Company, 1870.

Sixth Biennial Report of the Superintendent of Public Instruction for the Scholastic Years Ending August 31, 1887, and July 1, 1888, Being the Thirteenth Report from the Department of Education. Austin: State Printing Office, 1888.

Sixth Biennial Report of the Superintendent of Public Instruction of the State of Colorado, for Biennial Term Ending June 30, 1888. Denver: Collier and Cleaveland, 1889.

Superintendent of Public Instruction. *School Laws of Indiana, as Amended to March 15, 1877, with Opinions, Instructions and Judicial Decisions*

Relating to Common Schools and the Officers Thereof Indianapolis: Sentinel Company, 1877.

Texas Educational Survey Commission. *Texas Educational Survey Report: Courses of Study and Instruction.* Austin: Texas Educational Survey Commission, 1924.

Third Annual Report of the City Superintendent of Schools for the Year Ending July 31, 1901. New York: J. W. Pratt Company, n.d.

[*Third Biennial Report of the Superintendent of Public Instruction to the Governor of North Dakota, 1892–1894*]. N.p., n.d.

Thirteenth Biennial Report of the Superintendent of Public Instruction of the State of Illinois: October 1, 1878–June 30, 1880. Springfield: H. W. Rokker, 1881.

Thirty-Ninth Annual Report of the Superintendent of Public Instruction of the State of Michigan, with Accompanying Documents, for the Year 1875. Lansing: W. S. George and Co., 1876.

Thirty-Third Annual Report of the Board of President and Directors of the St. Louis Public Schools for the Year Ending June 30, 1887. St. Louis: Nixon-Jones Publishing Co., 1888.

Twelfth Annual Report of the Board of Education, Together with the Twelfth Annual Report of the Secretary of the Board. Boston: Dutton and Wentworth, 1849.

Twenty-Fifth Annual Report of the Board of Directors of the St. Louis Public Schools, for the Year Ending August 1, 1879. St. Louis: G. I. Jones and Co., 1880.

Twenty-First Annual Report of the Board of Directors of the St. Louis Public Schools, for the Year Ending August 1, 1875. St. Louis: Globe-Democrat Job Printing Co., 1876.

Twenty-Second Annual Report of the Public Schools of the City of Indianapolis, for the School Year Ending June 30, 1883. Indianapolis: Sentinel Co., 1883.

Twenty-Third Annual Report of the Superintendent of Public Instruction of the State of New York. Jerome B. Parmenter, State Printer, 1877.

Published Primary Sources

Addams, Jane. *Twenty Years at Hull-House, with Autobiographical Notes.* 1919. Reprint, with a foreword by Henry Steele Commager, New York: Signet Classic, 1961.

Antin, Mary. *The Promised Land*, 2d ed. Princeton: Princeton University Press, 1969.

Barnard, Henry, ed. "Subjects and Courses of Instruction in City Public Schools." *The American Journal of Education* 3 (1870): 465–576.

Barnard, Henry. "William Torrey Harris and St. Louis Public Schools." *American Journal of Education* 30 (1880): 625–641.

Bartlett, John Russell. *Dictionary of Americanisms: A Glossary of Word and Phrases, Usually Regarded as Peculiar to the United States.* 1848. Reprint, with a new introduction by Richard Lederer, Hoboken, NJ: John Wiley and Sons, Inc., 2003.

"Benz Explains Vote." *Indianapolis Star*, February 26, 1919.

"Berlin Sees a Wilson Trick." *New York Times*, March 6, 1917.

Bierstadt, Edward Hale. *Aspects of Americanization*. Cincinnati: Stewart Kidd Company, 1922.

"Bilingual Material in Libraries Draws Some Criticism." *New York Times*, September 5, 2005.

Boernstein, Henry. *Memoirs of a Nobody: The Missouri Years of an Austrian Radical, 1849–1866*. Translated and edited by Steven Rowan. St. Louis: Missouri Historical Society, 1997.

Brinton, Daniel G. *Races and Peoples: Lectures on the Science of Ethnography*. New York: N. D. C. Hodges, 1890.

Clark, Ann Nolan. *In My Mother's House*. 1941. Reprint, New York: Viking Press, 1966.

Clark, Dolly Holliday. "Memoir of a Country Schoolteacher: Dolly Holliday Meets the Ethnic West, 1919–1920." Edited by Paula M. Nelson. *North Dakota History 59*, no. 1 (1992): 30–45.

Comegys, Benjamin B., ed. *A Primer of Ethics*. Boston: Ginn and Company, 1891.

"Comment of This Morning's Newspapers on President Wilson's Announcement." *New York Times*, February 4, 1917.

Conant, James B. *The American High School Today: A First Report to Interested Citizens*. New York: McGraw-Hill Book Company, Inc., 1960.

Coulomb, Charles A., Armand J. Gerson, and Albert E. McKinley. *Outline of an Emergency Course of Instruction on the War*. Washington, DC: Government Printing Office, 1918.

Counts, George S. *Dare the School Build a New Social Order?*. 1932. Reprint, with a new preface by Wayne J. Urban, Carbondale: Southern Illinois University Press, 1978.

Covello, Leonard, with Guido D'Agostino. *The Heart Is the Teacher*. New York: McGraw-Hill, 1958.

Covello, Leonard. "Language as a Factor in Integration and Assimilation: The Role of the Language Teacher in a School-Community Program." *The Modern Language Journal 23*, no. 5 (1939): 323–333.

Cowan, Paul. *The Tribes of America: Journalistic Discoveries of Our People and Their Cultures*. New York: New Press, 2008.

Cubberley, Ellwood P. *Public Education in the United States: A Study and Interpretation of American Educational History*. Boston: Houghton Mifflin Company, 1919.

Dana, Richard Henry. *Two Years before the Mast*. Edited by Thomas Philbrick. New York: Penguin Books, 1981.

Darwin, Charles. *The Origin of Species*. 1859. Reprint, with a foreword by Patricia Horan, New York: Gramercy Books, 1979.

DeBow, J. D. *The Seventh Census of the United States: 1850*. Washington, DC: Robert Armstrong, 1853.

Debs, Theodore. *Sidelights: Incidents in the Life of Eugene V. Debs*. Terre Haute, IN: Moore-Langen Printing Co., N.d.

Department of the Interior, Census Office. *Statistics of the Population of the United States at the Tenth Census (June 1, 1880)*. Washington, DC: Government Printing Office, 1883.

Dewey, John. *Democracy and Education: An Introduction to the Philosophy of Education.* 1916. Reprint, New York: Free Press, 1997.

———. *Experience and Education.* 1938. Reprint, New York: Touchstone, 1997.

Dobbs, Lou. "English-Only Advocates See Barriers to Bill Easing Up." Available at www.cnn.com/2005/US/04/18/official.english/index.html.

Douai, Adolf, and John Straubenmueller. "German Schools in the United States." *The American Journal of Education* 3 (1870): 581–586.

Dougherty, N. C. "Recent Legislation upon Compulsory Education in Illinois and Wisconsin." In *Journal of Proceedings and Addresses,* National Educational Association, 393–401. New York: National Education Association, 1891.

Douglass, Frederick. *Narrative of the Life of Frederick Douglass: An American Slave.* New York: Signet Classic, 1997.

Du Bois, W. E. B. *The Souls of Black Folk.* 1903. Reprint, with a new introduction by Randall Kenan, New York: Signet Classic, 1995.

Edwards, I. N. "The Legal Status of Foreign Languages in the Schools." *The Elementary School Journal* 24, no. 4 (1923): 270–278.

Eliot, Charles W. "Address of Welcome." *Publications of the Modern Language Association of America* 5, no. 1 (1890): 1–4.

Emmerich, C. E. "No Such Sentiment Uttered." *Indianapolis News.* May 3, 1890.

Esquibel Tywoniak, Frances, and Mario T. García. *Migrant Daughter: Coming of Age as a Mexican American Woman.* Berkley: University of California Press, 2000.

"Esther Vaagen." *North Dakota History* 44, no. 4 (1977): 14–16.

Ettinger, William L. *Ten Addresses Delivered before Associate and District Superintendents of the New York City Schools and Other Professional Bodies.* N.p., n.d.

Fitz, Asa. *The American School Hymn Book.* Boston: Crosby, Nichols and Company, 1854.

Fletcher, Henry J., ed., "Recent Legislation Forbidding Teaching of Foreign Languages in Public Schools." *Minnesota Law Review* 4, no. 6 (1920): 449–551.

"For Freedom and Civilization." *New York Times,* April 3, 1917.

"Foreign Languages in the Elementary School." *School and Society* 6, no. 151 (1917): 583–584.

"German Schools." *New York Times.* December 8, 1889.

Gladden, Washington. "The Anti-Catholic Crusade." *Century* 47 (March 1894): 789–795.

Goodman, Paul. *Compulsory Mis-education.* New York: Horizon Press, 1964.

Grant, Percy Stickney. "American Ideals and Race Mixture." *North American Review* 195, no. 4 (1912): 513–525.

Hall, G. Stanley. "The White Man's Burden Versus Indigenous Development for the Lower Races." In *Journal of Proceedings and Addresses of the National Educational Association,* National Educational Association, 1053–1056. Chicago: University of Chicago Press, 1903.

Hall, Prescott F. "The Future of American Ideals." *North American Review* 195, no. 1 (1912): 94–102.

Hall, Prescott F. "Immigration and the World War." *Annals of the American Academy of Political and Social Science* 93 (January 1921): 190–193.

———. "Italian Immigration." *North American Review* 163, no. 2 (1896): 252–254.

Handschin, Charles Hart. *The Teaching of Modern Languages in the United States.* Washington, DC: Government Printing Office, 1913.

Hennighausen, Louis P. "Reminiscences of the Political Life of the German-Americans in Baltimore during the Years 1850–1860." In *Eleventh and Twelfth Annual Reports of the Society for the History of the Germans in Maryland, 1897–1898,* 3–18. N.p., n.d.

Hill, David Jayne. "Dual Citizenship in the German Imperial and State Citizenship Law." *American Journal of International Law* 12 (April 1918): 356–363.

Hinsdale, Mary L. "A Legislative History of the Public School System of the State of Ohio." In *Annual Reports of the Department of Interior for the Fiscal Year Ended June 30, 1901: Report of the Commission of Education,* 129–159. Vol. 1. Washington, DC: Government Printing Office, 1902.

Hoffmann, Else. "The German-Language School in Baltimore." *The German-American Review* 5, no. 2 (1938): 34–35, 55.

"House Passes M'Cray's Bill in 15 Minutes." *Indianapolis Star,* February 26, 1919.

Howard C. Hill. "The Americanization Movement." *The American Journal of Sociology* 24, no. 6 (1919): 609–642.

Institute for Government Research. *The Problem of Indian Administration.* Baltimore: Johns Hopkins Press, 1928.

Jordan, David Starr. "Nature Study and Moral Culture." *Science* 4, no. 84 (1896): 149–156.

———. *The Value of Higher Education: An Address to Young People.* Richmond, IN: Daily Palladium Book and Job Printing House, 1888.

Kennedy, Joseph C. G. *Population of the United States in 1860; Compiled from the Original Returns of the Eighth Census.* Washington, DC: Government Printing Office, 1864.

Klemm, L. R. *Public Education in Germany and the United States.* Boston: Gorham Press, 1911.

Kroeh, Charles F. "Methods of Teaching Modern Languages." *Transactions and Proceedings of the Modern Language Association of America* 3 (1887): 169–185.

Kroeh, Charles F., H. C. G. von Jagemann, L. A. Staeger, O. Seidensticker, Paul F. Rohrbacher, and C. Sprague Smith. "Methods of Teaching Modern Languages: Discussion." *Transactions and Proceedings of the Modern Language Association of America* 3 (1887): xxxiii–xxxvi.

La Flesche, Francis. *The Middle Five: Indian Schoolboys of the Omaha Tribe.* 1900. Reprint, with an introduction by David A. Baerreis, Madison: University of Wisconsin Press, 1963.

Learned, M. D. "German in the Public Schools." *German American Annals,* vol. 2 (1913):100–106.

"Letter from a Rabid Know-Nothing—Views, Feelings, and Aims of the Party." *New York Daily Times.* June 8, 1854.

López-Stafford, Gloria. *A Place in El Paso: A Mexican-American Childhood.* Albuquerque: University of New Mexico Press, 1996.

Manuel, Herschel T. *The Education of Mexican and Spanish-Speaking Children in Texas.* Austin: University of Texas, 1930.

Marx, Karl, and Frederick Engels. *The Communist Manifesto.* New York: International Publishers, 1997.

Medina, Pablo. *Exiled Memories: A Cuban Childhood.* New York: Persea Books, 2002.

Meyer v. Nebraska, 262 U.S. 390 (1923).

Mills, W. W. *Forty Years at El Paso, 1858–1898.* El Paso: Carl Hertzog, 1962.

Morris, J. G. "The German in Baltimore." In *Eighth, Ninth and Tenth Annual Reports of the Society for the History of the Germans in Maryland, 1894–1896*, 11–19. N.p., n.d.

"National German-American Teachers' Association." In *Report of the Secretary of the Interior; Being Part of the Message and Documents Communicated to the Two Houses of Congress at the Beginning of the Third Session of the Forty-Sixth Congress*, 397–398. Vol. 3. Washington, DC: Government Printing Office, 1882.

Nava, Julian. *My Mexican-American Journey.* Houston: Arte Público Press, 2002.

Ohlinger, Gustavus. *The German Conspiracy in American Education.* New York: George H. Doran Company, [1919?].

———. *Their True Faith and Allegiance.* New York: Macmillan Company, 1916.

Parkman, Francis. *The Oregon Trail.* Edited by Mason Wade. Heritage Press, 1943.

Peaslee, John B. *Thoughts and Experiences in and out of School.* Cincinnati: Curts and Jennings, 1900.

Pelton, John Cotter. *Life's Sunbeams and Shadows: Poems and Prose with Appendix, Including Biographical and Historical Notes in Prose.* Vol. 1. San Francisco: Bancroft Company, 1893.

Pierce v. Society of Sisters, 268 U.S. 510 (1925).

Pratt, Richard Henry. *Battlefield and Classroom: Four Decades with the American Indian, 1867–1904.* Edited by Robert M. Utley. New Haven: Yale University Press, 1964.

"Quick Action Demanded." *Indianapolis Star*, February 26, 1919.

Report of the Secretary of the Interior, Being Part of the Message and Documents Communicated to the Two Houses of Congress at the Beginning of the Third Session of the Forty-First Congress. Vol. 2. Washington, DC: Government Printing Office, 1870.

Report of the Secretary of the Interior, Being Part of the Message and Documents Communicated to the Two Houses of Congress at the Beginning of the Second Session of the Forty-Third Congress. Vol. 2. Washington, DC: Government Printing Office, 1875.

Report of the Secretary of the Interior; Being Part of the Message and Documents Communicated to the Two Houses of Congress at the Beginning of the Second Session of the Forty-Sixth Congress. Vol. 1. Washington, DC: Government Printing Office, 1879.

Report of the Secretary of the Interior; Being Part of the Message and Documents Communicated to the Two Houses of Congress at the Beginning of the Second Session of the Forty-Eighth Congress. Vol. 2. Washington, DC: Government Printing Office, 1884.

Report of the Secretary of the Interior; Being Part of the Message and Documents Communicated to the Two Houses of Congress at the Beginning of the First Session of the Fiftieth Congress. Vol. 2. Washington, DC: Government Printing Office, 1887.

Reynolds, Annie. *The Education of Spanish-Speaking Children in Five Southwestern States.* Washington, DC: Government Printing Office, 1933.

Rickover, H. G. *American Education—A National Failure: The Problem of Our Schools and What We Can Learn from England.* New York: E. P. Dutton and Co., Inc., 1963.

Riis, Jacob A. *How the Other Half Lives.* 1890. Reprint, with an introduction by Francesco Cordasco, New York: Garrett Press, Inc., 1970.

Roeming, Robert F. "Bilingualism and the National Interest." *The Modern Language Journal* 55, no. 2 (1971): 73–81.

"Roosevelt Demands Speed-Up of War." *New York Times,* August 27, 1918.

Rosenfeld, Jessie. "Special Classes in the Public Schools of New York." *Education* 27, no. 2 (1906): 92–100.

Sanchez, George Isidore. "The Education of Bilinguals in a State School System." PhD diss., University of California, 1934.

Schrader, Frederick Franklin. *Handbook Political, Statistical and Sociological for German Americans and All Other Americans Who Have Not Forgotten the History and Traditions of Their Country and Who Believe in the Principles of Washington, Jefferson and Lincoln.* New York: n.p., 1916–1917.

Standing Bear, Luther. *My People the Sioux.* Edited by E. A. Brininstool. 1928. Reprint, with an introduction by Richard N. Ellis, Lincoln: University of Nebraska Press, 1975.

Steffen, Wm. "German Schools and Teaching German." In *Report of the Secretary of the Interior, Being Part of the Message and Documents Communicated to the Two Houses of Congress at the Beginning of the Third Session of the Forty-First Congress,* 437–438. Vol. 2. Washington, DC: Government Printing Office, 1870.

Stempfel, Theodore. *Fifty Years of Unrelenting German Aspirations in Indianapolis.* 1898. Reprint, edited by Giles R. Hoyt, Claudia Grossmann, Elfrieda Lang, and Eberhard Reichmann, Indianapolis: German-American Center and Indiana German Heritage Society, Inc., 1991.

Swarns, Rachel L. "Immigrants Rally in Scores of Cities for Legal Status." *New York Times,* April 11, 2006.

———. "Senate Deal on Immigration Falters." *New York Times,* April 7, 2006.

Swett, John. *Public Education in California.* 1911. Reprint, New York: Arno Press, 1969.

Thompson, Frank V. *Schooling of the Immigrant.* New York: Harper and Brothers Publishers, 1920.

"To Our Readers." *Telegraph and Tribüne* (Indianapolis). May 31, 1918.

Tocqueville, Alexis de. *Democracy in America and Two Essays on America.* Edited by Isaac Kramnick. Translated by Gerald E. Bevan. New York: Penguin Books, 2003.

U.S. Bureau of Education. *Education of the Immigrant.* Washington, DC: Government Printing Office, 1913.

———. *Education in Patriotism: A Synopsis of the Agencies at Work.* Teachers' Leaflet, no. 2, 1918.

———. *Government Policies Involving the Schools in War Time.* Teachers' Leaflet, no. 3, 1918.

———. *A Survey of Education in Hawaii.* Washington, DC: Government Printing Office, 1920.

U.S. Children's Bureau. *Baby-Saving Campaigns: A Preliminary Report on What American Cities Are Doing to Prevent Infant Mortality.* Washington, DC: Government Printing Office, 1913.

U.S. Congress. *An Address to the People of the United States upon the Social, Moral, and Political Effect of Chinese Immigration.* 45th Cong., 1st sess., 1877, H. Mis. Doc. 9.

———. *Chinese Immigration.* 45th Cong., 2d sess., 1878, H. Rep. 240.

U.S. Immigration Commission. *The Children of Immigrants in Schools.* Vol. 1. Washington, DC: Government Printing Office, 1911.

———. *Dictionary of Races or Peoples.* Washington, DC: Government Printing Office, 1911.

———. *Statistical Review of Immigration, 1820–1910—Distribution of Immigrants, 1850–1900.* Washington, DC: Government Printing Office, 1911.

"Varia: French in the Public Elementary Schools in Louisiana." *The French Review* 13, no. 4 (1940): 344–346.

Viereck, L. "German Instruction in American Schools." In *Annual Reports of the Department of Interior for the Fiscal Year Ended June 30, 1901: Report of the Commission of Education,* 531–708. Vol. 1. Washington, DC: Government Printing Office, 1902.

Vonnegut, Clemens. *A Proposed Guide for Instruction in Morals from the Standpoint of a Freethinker, for Adult Persons Offered by a Dilettante.* 1900. Reprint, n.p., 1987.

Walker, Francis A. *The Statistics of the Population of the United States, Embracing the Tables of Race, Nationality, Sex, Selected Ages, and Occupations.* Ninth Census. Vol. 1. Washington, DC: Government Printing Office, 1872.

Washington, Booker T. *Up From Slavery.* 1901. Reprint, with a new introduction by Ishmael Reed, New York: Signet Classic, 2000.

West, Mrs. Max. *Infant Care.* Washington, DC: Government Printing Office, 1914.

Wilkins, Lawrence A. "Spanish as a Substitute for German for Training and Culture." *Hispania* 1, no. 4 (1918): 205–221.

Willson, Marcius. *The Fourth Reader of the United States Series.* New York: Harper and Brothers, 1872.

Zehr, Mary Ann. "English-Only Advocate Uses Ariz. State Office to Carry out Mission." *Education Week,* February 1, 2006.

Edited Collections of Primary Sources

Addams, Jane. *On Education.* Edited by Ellen Condliffe Lagemann. New Brunswick: Transaction Publishers, 2002.

Archambault, Reginald D., ed. *John Dewey on Education: Selected Writings.* Chicago: University of Chicago Press, 1974.

Bremner, Robert H., ed. *Children and Youth in America: A Documentary History.* 2 vols. Cambridge, MA: Harvard University Press, 1971.

Bruun, Erik, and Jay Crosby, eds. *Our Nation's Archives: The History of the United States in Documents.* New York: Tess Press, 1999.

Cordasco, Francesco, ed. *The Italian Community and Its Language in the United States: The Annual Reports of the Italian Teachers Association.* Totowa, NJ: Rowman and Littlefield, 1975.

Crawford, James, ed. *Language Loyalties: A Source Book on the Official English Controversy.* Chicago: University of Chicago Press, 1992.

Fraser, James W., ed. *The School in the United States: A Documentary History.* New York: McGraw-Hill, 2001.

Handlin, Oscar, ed. *Immigration as a Factor in American History.* Englewood Cliffs, NJ: Prentice-Hall Inc., 1959.

Hoffman, Nancy, ed. *Woman's "True" Profession: Voices from the History of Teaching.* Old Westbury, NY: Feminist Press, 1981.

Hofstadter, Richard, ed. *The Progressive Movement, 1900–1915.* New York: Touchstone, 1986.

Hollinger, David A., and Charles Capper, eds. *The American Intellectual Tradition: A Sourcebook.* Vol. 2. 4th ed. New York: Oxford University Press, 2001.

James, William. *Essays in Pragmatism.* Edited by Alburey Castell. New York: Hafner Press, 1948.

Joshi, S. T., ed. *Documents of American Prejudice: An Anthology of Writings on Race from Thomas Jefferson to David Duke.* New York: Basic Books, 1999.

Kamphoefner, Walter D., Wolfgang Helbich, and Ulrike Sommer, eds. *News from the Land of Freedom: German Immigrants Write Home.* Translated by Susan Carter Vogel. Ithaca: Cornell University Press, 1991.

Lixl-Purcell, Andreas, ed. *Stimmen eines Jahrhunderts 1888–1990: Deutsche Autobiographien, Tagebücher, Bilder und Briefe.* Fort Worth: Holt, Rinehart and Winston, Inc., 1990.

Miller, Tom, ed. *How I Learned English: 55 Accomplished Latinos Recall Lesson in Language and Life.* Washington, DC: National Geographic Society, 2007.

O'Neill, Teresa. *Immigration: Opposing Viewpoints.* San Diego: Greenhaven Press, Inc., 1992.

Prucha, Francis Paul, ed. *Documents of United States Indian Policy.* 3rd ed. Lincoln: University of Nebraska Press, 2000.

Weber, David J., ed. *Foreigners in Their Native Land: Historical Roots of the Mexican Americans.* 30th anniversary ed. Albuquerque: University of New Mexico Press, 2003.

Zempel, Solveig, ed. *In Their Own Words: Letters from Norwegian Immigrants.* Translated by Solveig Zempel. Minneapolis: University of Minnesota Press, 1991.

Literature

Alger, Jr., Horatio. *Ragged Dick: Or, Street Life in New York with the Boot-blacks*. 1868. Reprint, with an introduction by Alan Trachtenberg, New York: Signet Classic, 1990.

Aurelius, Marcus. *Meditations*. Translated by Maxwell Staniforth. Baltimore: Penguin Books, 1964.

Bellow, Saul. *The Adventures of Augie March*. New York: Penguin Books, 1996.

Boernstein, Henry. *The Mysteries of St. Louis*. Translated by Friedrich Münch. 1851. Modern edition by Steven Rowan and Elizabeth Sims, Chicago: Charles H. Kerr Publishing Company, 1990.

Burroughs, William S. *My Education: A Book of Dreams*. New York: Penguin Books, 1996.

Carter, Forrest. *The Education of Little Tree*. Albuquerque: University of New Mexico Press, 1976.

Cather, Willa. *Death Comes for the Archbishop*. 1925. Reprint, New York: Vintage Classics, 1990.

————. *My Ántonia*. 1918. Reprint, Boston: Houghton Mifflin, 1954.

Céline, Louis-Ferdinand. *Journey to the End of the Night*. Translated by Ralph Manheim. New York: New Directions, 2006.

Cisneros, Sandra. *The House on Mango Street*. New York: Vintage Contemporaries, 1991.

Crane, Stephen. *The Blue Hotel*. Edited by Joseph Katz. Columbus: Charles E. Merrill Publishing Company, 1969.

Dos Passos, John. *1919*. New York: Harcourt, Brace and Company, 1932.

Dreiser, Theodore. *Dawn*. New York: Horace Liveright, Inc., 1931.

————. *A Hoosier Holiday*. 1916. Reprint, with a new introduction by Douglas Brinkley, Bloomington: Indiana University Press, 1998.

Eggleston, Edward. *The Hoosier School-Master*. 1871. Reprint, with and introduction by Vernon Loggins, New York: Hill and Wang, 1957.

Ellison, Ralph. *Invisible Man*. New York: Vintage International, 1995.

Fast, Howard. *The Immigrants*. Boston: Houghton Mifflin Company, 1977.

Kundera, Milan. *Ignorance*. Translated by Linda Asher. New York: Perennial, 2002.

Lewis, Sinclair. *Babbitt*. 1922. Reprint, Mineola, NY: Dover Publications, Inc., 2003.

Márquez, Gabriel García. *One Hundred Years of Solitude*. Translated by Gregory Rabassa. New York: Perennial Classics, 1998.

McCarthy, Cormac. *Blood Meridian, or the Evening Redness in the West*. New York: Modern Library, 2001.

Melville, Herman. *The Confidence Man*. New York: Prometheus Books, 1995.

Morrison, Toni. *The Bluest Eye*. New York: Vintage International, 2007.

Pamuk, Orhan. *Snow*. Translated by Maureen Freely. New York: Vintage International, 2005.

Pears, Iain. *An Instance of the Fingerpost*. New York: Riverhead Books, 2000.

Plath, Sylvia. *The Bell Jar*. New York: Perennial Classics, 1999.

Rolvaag, O. E. *Giants in the Earth.* New York: Harper and Brothers Publishers, 1929.

Sinclair, Upton. *The Jungle.* 1906. Reprint, with an introduction by Ronald Gottesman. New York: Penguin Books, 1986.

Twain, Mark. *Life on the Mississippi.* Mineola, NY: Dover Publications, 2000.

Vonnegut, Kurt. *Bagombo Snuff Box.* New York: Berkley Books, 2000.

———. *Bluebeard.* New York: Dial Press, 2006.

———. *God Bless You, Mr. Rosewater.* New York: Delta, 1998.

———. *A Man without a Country.* New York: Seven Stories Press, 2005.

———. *Palm Sunday: An Autobiographical Collage.* New York: Delta, 1999.

Woolf, Virginia. *Jacob's Room.* 1922. Reprint, with an introduction by Danell Jones, New York: Barnes & Noble, 2006.

Secondary Sources

Adams, David Wallace. *Education for Extinction: American Indians and the Boarding School Experience, 1875–1928.* Lawrence: University Press of Kansas, 1995.

———. "Land, Law, and Education: The Troubled History of Indian Citizenship, 1871–1924." In *Civic and Moral Learning in America*, edited by Donald Warren and John J. Patrick, 119–34. New York: Palgrave Macmillan, 2006.

Anderson, Arlow W. *Rough Road to Glory: The Norwegian-American Press Speaks out on Public Affairs, 1875–1925.* Philadelphia: Balch Institute, 1990.

Anderson, Benedict. *Imagined Communities: Reflections on the Origin and Spread of Nationalism.* Revised ed. New York: Verso, 1991.

Anderson, James D. *The Education of Blacks in the South, 1860–1935.* Chapel Hill: University of North Carolina Press, 1988.

Andersson, Theodore. "Bilingual Education: The American Experience." *The Modern Language Journal* 55, no. 7 (1971): 427–40.

Andersson, Theodore, and Mildred Boyer. *Bilingual Schooling in the United States: History, Rationale, Implications, and Planning.* Vol. 1. Austin: Southwest Educational Development Laboratory, 1970.

Angus, David L., and Jeffrey E. Mirel. *The Failed Promise of the American High School, 1890–1995.* New York: Teachers College Press, 1999.

———. "Presidents, Professors, and Lay Boards of Education: The Struggle for Influence over the American High School, 1860–1910." In *A Faithful Mirror: Reflections on the College Board and Education in America*, edited by Michael C. Johanek, 3–42. New York: College Entrance Examination Board, 2001.

Asato, Noriko. *Teaching Mikadoism: The Attack on Japanese Language Schools in Hawaii, California, and Washington, 1919–1927.* Honolulu: University of Hawai'i Press, 2006.

Ashby, LeRoy. *Saving the Waifs: Reformers and Dependent Children, 1890–1917.* Philadelphia: Temple University Press, 1984.

Atlas of American History. Skokie, IL: Houghton Mifflin / Rand McNally, 1993.

Bailyn, Bernard. *Education and the Forming of American Society: Needs and Opportunities for Study.* New York: W. W. Norton and Company, 1972.

Beard, Charles A. *Economic Origins of Jeffersonian Democracy.* New York: Free Press, 1965.

Becker, Ernest J. "History of the English-German Schools in Baltimore." In *Twenty-Fifth Report, Society for the History of the Germans in Maryland,* 13–17. Baltimore: N.p., 1942.

Bender, Thomas. *Toward an Urban Vision: Ideas and Institutions in Nineteenth-Century America.* Lexington: University Press of Kentucky, 1975.

Bennett, William J. *The De-Valuing of America: The Fight for Our Culture and Children.* New York: Summit, 1992.

Benseler, David P. "In Memoriam: Robert F. Roeming, 1912–2004." *The Modern Language Journal* 89, no. 2 (2005): 159–160.

Berrol, Selma C. *Julia Richman: A Notable Woman.* Philadelphia: Balch Institute Press, 1993.

Berrol, Selma Cantor. *Growing up American: Immigrant Children in America, Then and Now.* New York: Twayne Publishers, 1995.

———. *Immigrants at School: New York City, 1898–1914.* New York: Arno Press, 1978.

Blanton, Carlos Kevin. "The Rise of English-Only Pedagogy: Immigrant Children, Progressive Education, and Language Policy in the United States, 1900–1930." In *When Science Encounters the Child: Education, Parenting, and Child Welfare in 20th-Century America,* edited by Barbara Beatty, Emily D. Cahan, and Julia Grant, 56–76. New York: Teachers College Press, 2006.

———. *The Strange Career of Bilingual Education in Texas, 1836–1981.* College Station: Texas A&M University Press, 2004.

Bodnar, John. *The Transplanted: A History of Immigrants in Urban America.* Bloomington: Indiana University Press, 1985.

Brumberg, Stephan F. *Going to America, Going to School: The Jewish Immigrant Public School Encounter in Turn-of-the-Century New York City.* New York: Praeger, 1986.

Bull, Barry L., Royal T. Fruehling, and Vergie Chattergy. *The Ethics of Multicultural and Bilingual Education.* New York: Teachers College Press, 1992.

Burnham, Walter Dean. "Table I: Summary: Presidential Elections, USA, 1788–2004." *The Journal of the Historical Society* 7, no. 4 (2007): 521–580.

Cahan, Emily D. "Toward a Socially Relevant Science: Notes on the History of Child Development Research." In *When Science Encounters the Child: Education, Parenting, and Child Welfare in 20th-Century America,* edited by Barbara Beatty, Emily D. Cahan, and Julia Grant, 16–34. New York: Teachers College Press, 2006.

Carlson, Robert A. *The Quest for Conformity: Americanization through Education.* New York: John Wiley and Sons, Inc., 1975.

Castellanos, Diego. *The Best of Two Worlds: Bilingual-Bicultural Education in the U.S.* 2nd ed. Trenton: New Jersey State Department of Education, 1985.

Chambers, Theodore Frelinghuysen. *The Early Germans of New Jersey: Their Histories, Churches and Genealogies.* Baltimore: Genealogical Publishing Company, 1969.

Chomsky, Noam. *Media Control: The Spectacular Achievements of Propaganda.* 2nd ed. New York: Seven Stories Press, 2002.

Christian, Donna, and Fred Genesee. *Bilingual Education.* Alexandria, VA: TESOL, 2001.

Clark, Duncan. *A New World: The History of Immigration into the United States.* San Diego: Thunder Bay Press, 2000.

Clark, George S. "The German Presence in New Jersey." In *The New Jersey Ethnic Experience,* edited by Barbara Cunningham, 219–28. Union City, NJ: Wm. H. Wise and Co., 1977.

Commager, Henry Steele. *The American Mind: An Interpretation of American Thought and Character since the 1880's.* New Haven: Yale University Press, 1950.

Cordasco, Francesco. *Bilingual Schooling in the United States: A Sourcebook for Educational Personnel.* New York: McGraw-Hill, 1976.

———, ed. *Dictionary of American Immigration History.* Metuchen, NJ: Scarecrow Press, 1990.

Crawford, James. *Bilingual Education: History, Politics, Theory, and Practice.* 3rd ed. Los Angeles: Bilingual Educational Services, 1995.

Cremin, Lawrence A. *American Education: The Colonial Experience, 1607–1783.* New York: Harper and Row, 1970.

———. *The Transformation of the School: Progressivism in American Education, 1876–1957.* New York: Vintage Books, 1964.

Cuban, Larry. *How Teachers Taught: Constancy and Change in American Classrooms, 1880–1990.* 2nd ed. New York: Teachers College Press, 1993.

Cunz, Dieter. *The Maryland Germans, a History.* Princeton: Princeton University Press, 1948.

Curti, Merle. "American Philanthropy and the National Character." *American Quarterly* 10 (1958): 420–37.

———. *The Social Ideas of American Educators.* Paterson, NJ: Littlefield, Adams and Co., 1959.

Davis, Matthew D. *Exposing a Culture of Neglect: Herschel T. Manuel and Mexican American Schooling.* Greenwich, CT: Information Age Publishing, 2005.

Detjen, David W. *The Germans in Missouri, 1900–1918: Prohibition, Neutrality, and Assimilation.* Columbia: University of Missouri Press, 1985.

Devore, Donald E., and Joseph Logsdon. *Crescent City Schools: Public Education in New Orleans, 1841–1991*: The Center for Louisiana Studies, University of Southwestern Louisiana, 1991.

Diner, Hasia R. "The World of Whiteness." *Historically Speaking* 9, no. 1 (2007): 20–22.

Donato, Ruben. "Hispano Education and the Implications of Autonomy: Four School Systems in Southern Colorado, 1920–1963." *Harvard Educational Review* 69, no. 2 (1999): 117–49.

———. *The Other Struggle for Equal Schools: Mexican Americans During the Civil Rights Era.* Albany: State University of New York Press, 1997.

Einstein, Albert and Sigmund Freud. *Why War?* Translated by Stuart Gilbert: International Institute of Intellectual Co-operation, League of Nations, 1933.

Ellis, Frances H. "German Instruction in the Public Schools of Indianapolis, 1869–1919, III." *Indiana Magazine of History* 50, no. 4 (1954): 357–80.

———. "Historical Account of German Instruction in the Public Schools of Indianapolis 1869–1919." *Indiana Magazine of History* 50, no. 2 (1954): 119–38.

Ellis, Joseph J. *Founding Brothers: The Revolutionary Generation.* New York: Vintage Books, 2002.

Elson, Ruth Miller. *Guardians of Tradition: American Schoolbooks of the Nineteenth Century.* Lincoln: University of Nebraska Press, 1964.

Epstein, Noel. *Language, Ethnicity, and the Schools: Policy Alternatives for Bilingual-Bicultural Education.* Washington, DC: Institute for Educational Leadership, 1977.

Evans, Sara M. *Born for Liberty: A History of Women in America.* New York: Free Press, 1991.

Ewen, Stuart, and Elizabeth Ewen. *Typecasting: On the Arts and Sciences of Human Inequality.* New York: Seven Stories Press, 2006.

Fass, Paula S. *Children of a New World: Society, Culture, and Globalization.* New York: New York University Press, 2007.

———. *Outside In: Minorities and the Transformation of American Education.* New York: Oxford University Press, 1989.

Fessler, Paul Rudolph. "Speaking in Tongues: German-Americans and the Heritage of Bilingual Education in American Public Schools." PhD diss., Texas A & M University, 1997.

Fishman, Joshua A. *Hungarian Language Maintenance in the United States.* Bloomington: Indiana University, 1966.

———. *Language Loyalty in the United States: The Maintenance and Perpetuation of Non-English Mother Tongues by American Ethnic and Religious Groups.* London: Mouton & Co., 1966.

Fogel, Robert William, and Stanley L. Engerman. *Time on the Cross: The Economics of American Negro Slavery.* New York: W. W. Norton and Company, 1995.

Foucault, Michel. *The Archaeology of Knowledge and the Discourse on Language.* Translated by A. M. Sheridan Smith. New York: Pantheon Books, 1972.

Fraser, James W. *Between Church and State: Religion and Public Education in a Multicultural America.* New York: St. Martin's Griffin, 2000.

Freud, Sigmund. *Civilization and Its Discontents.* Translated by James Strachey. New York: W. W. Norton and Company, 1989.

Friedman, Lawrence J. *Inventors of the Promised Land.* New York: Alfred A. Knopf, 1975.

Fromm, Erich. *Escape from Freedom.* New York: Henry Holt and Company, 1994.

García, Eugene E. *Teaching and Learning in Two Languages: Bilingualism and Schooling in the United States.* New York: Teachers College Press, 2005.

Gerber, David A. "Language Maintenance, Ethnic Group Formation, and Public Schools: Changing Patterns of German Concern, Buffalo, 1837–1874." *Journal of American Ethnic History* 4, no. 1 (1984): 31–61.

Gere, Anne Ruggles. "Indian Heart/White Man's Head: Native-American Teachers in Indian Schools, 1880–1930." *History of Education Quarterly* 45, no. 1 (2005): 38–65.

Gerstle, Gary. *American Crucible: Race and Nation in the Twentieth Century.* Princeton: Princeton University Press, 2001.

Goldberg, Bettina. "The German-English Academy, the National German-American Teachers' Seminary, and the Public School System in Milwaukee, 1851–1919." In *German Influences on Education in the United States to 1917*, edited by Henry Geitz, Jürgen Heideking, and Jurgen Herbst, 177–92. Cambridge: Cambridge University Press, 1995.

Gonzales, Manuel G. *Mexicanos: A History of Mexicans in the United States,* 1999.

Greven, Philip. *The Protestant Temperament: Patterns of Child-Rearing, Religious Experience, and the Self in Early America.* Chicago: University of Chicago Press, 1988.

Hacsi, Timothy A. *Children as Pawns: The Politics of Educational Reform.* Cambridge, MA: Harvard University Press, 2002.

Hartmann, Edward George. *The Movement to Americanize the Immigrant.* New York: AMS Press, Inc., 1967.

Haugen, Einar. *The Norwegian Language in America: A Study in Bilingual Behavior.* 2 vols. Philadelphia: University of Pennsylvania Press, 1953.

Hawes, Joseph M. *Children in Urban Society: Juvenile Delinquency in Nineteenth-Century America.* New York: Oxford University Press, 1971.

Heath, Shirley Brice. "A National Language Academy?: Debate in the New Nation." *International Journal of the Sociology of Language* 11 (1976): 9–43.

Higham, John. "Changing Paradigms: The Collapse of Consensus History." *The Journal of American History* 76, no. 2 (1989): 460–66.

———. *Strangers in the Land: Patterns of American Nativism, 1860–1925.* New Brunswick: Rutgers University Press, 1955.

Hobsbawm, Eric. *The Age of Capital, 1848–1875.* New York: Vintage Books, 1996.

———. *The Age of Revolution, 1789–1848.* New York: Vintage Books, 1996.

Hoffmann, Else. "The German-Language School in Baltimore." *The American-German Review* 5, no. 2 (1938): 34–35, 55.

Hofstadter, Richard. *The Age of Reform: From Bryan to F. D. R.* New York: Vintage Books, 1955.

———. *Social Darwinism in American Thought.* Boston: Beacon Press, 1992.

Hogan, David. "Education and the Making of the Chicago Working Class." *History of Education Quarterly* 18, no. 3 (1978): 227–70.

Hollinger, David A. *Postethnic America: Beyond Multiculturalism.* New York: Basic Books, 1995.

Hoyt, Giles R. "Germans." In *Peopling Indiana: The Ethnic Experience*, edited by Robert M. Taylor and Connie A. McBirney, 146–81. Indianapolis: Indiana Historical Society, 1996.

Ignatiev, Noel. *How the Irish Became White.* New York: Routledge Classics, 2009.

Jacobi, Juliane. "Schoolmarm, *Volkserzieher, Kantor,* and *Schulschwester*: German Teachers among Immigrants During the Second Half of the Nineteenth Century." In *German Influences on Education in the United States to 1917,* edited by Henry Geitz, Jürgen Heideking, and Jurgen Herbst, 115–28. Cambridge: Cambridge University Press, 1995.

Jordan, Winthrop D. *The Whie Man's Burden: Historical Origins of Racism in the United States.* London: Oxford University Press, 1974.

Jorgenson, Lloyd P. *The State and the Non-Public School, 1825–1925.* Columbia: University of Missouri Press, 1987.

Kaestle, Carl F. *Pillars of the Republic: Common Schools and American Society, 1780–1860.* New York: Hill and Wang, 2001.

Kamphoefner, Walter D. "German American Bilingualism: *cui malo?* Mother Tongue and Socioeconomic Status among the Second Generation in 1940." *International Migration Review* 28, no. 4 (1994): 846–864.

———. "Learning from the 'Majority-Minority' City: Immigration in Nineteenth-Century St. Louis." In *St. Louis in the Century of Henry Shaw: A View Beyond the Garden Wall,* edited by Eric Sandweiss, 79–99. Columbia: University of Missouri Press, 2003.

Katz, Michael B. *In the Shadow of the Poorhouse: A Social History of Welfare in America.* 10th anniversary ed. New York: Basic Books, 1996.

———. *The Irony of Early School Reform: Educational Innovation in Mid-Nineteenth Century Massachusetts.* New York: Teachers College, 2001.

Kazal, Russell A. *Becoming Old Stock: The Paradox of German-American Identity.* Princeton: Princeton University Press, 2004.

Keller, Morton. *Regulating a New Society: Public Policy and Social Change in America, 1900–1933.* Cambridge, MA: Harvard University Press, 1994.

Kennedy, David M. *Over Here: The First World War and American Society.* 25th anniversary ed. New York: Oxford University Press, 2004.

Kidd, Thomas S. "What Happened to the Puritans?" *Historically Speaking* 7, no. 1 (2005): 32–34.

Kinzer, Donald L. *An Episode of Anti-Catholicism: The American Protective Association.* Seattle: University of Washington Press, 1964.

Kirschbaum, Erik. *The Eradication of German Culture in the United States: 1917–1918.* Stuttgart: Hans-Dieter Heinz, 1986.

Kliebard, Herbert M. *Schooled to Work: Vocationalism and the American Curriculum, 1876–1946.* New York: Teachers College Press, 1999.

———. *The Struggle for the American Curriculum, 1893–1958.* 3rd ed. New York: RoutledgeFalmer, 2004.

Kloss, Heinz. *The American Bilingual Tradition.* 1977. Reprint, with a new introduction by Reynaldo F. Macias and Terrence G. Wiley, McHenry, IL: Center for Applied Linguistics and Delta Systems Co., 1998.

———. *Das Nationalitätenrecht Der Vereinigten Staaten Von Amerika.* Wien: Wilhelm Braumüller, 1963.

———, ed. *Laws and Legal Documents Relating to Problems of Bilingual Education in the United States.* Washington, DC: Center for Applied Linguistics, 1971.

Kozol, Jonathan. *The Shame of the Nation: The Restoration of Apartheid Schooling in America*. New York: Crown Publishers, 2005.

Lange, Dena. "Information Concerning One Hundred Years of Progress in the St. Louis Public Schools." *Public School Messenger* 35, no. 5 (1938).

Lasch, Christopher. *Haven in a Heartless World: The Family Besieged*. New York: Basic Books, 1977.

Lederman, Sara Henry. "Philanthropy and Social Case Work: Mary E. Richmond and the Russell Sage Foundation, 1909–1928." In *Women and Philanthropy in Education*, edited by Andrea Walton, 60–80. Bloomington: Indiana University Press, 2005.

Leibowitz, Arnold H. "A History of Language Policy in American Indian Schools." In *Bilingual Education for American Indians*, 1–6. New York: Arno Press, 1978.

Lomawaima, Tsianina K., and Teresa L. McCarty. *"To Remain an Indian": Lessons in Democracy from a Century of Native American Education*. New York: Teachers College Press, 2006.

Lovejoy, Arthur O. *The Great Chain of Being: A Study of the History of an Idea*. Cambridge, MA: Harvard University Press, 1936.

Lovoll, Odd S. *The Promise of America: A History of the Norwegian-American People*. Minneapolis: University of Minnesotsa Press, 1984.

Low, Victor. *The Unimpressible Race: A Century of Educational Struggle by the Chinese in San Francisco*. San Francisco: East/West Publishing Company, 1982.

Luebke, Frederick C. *Bonds of Loyalty: German-Americans and World War I*. DeKalb, IL: Northern Illinois University Press, 1974.

———. "Ethnic Group Settlement on the Great Plains." *The Western Historical Quarterly* 8, no. 4 (1977): 405–30.

———. "The German-American Alliance in Nebraska, 1910–1917." *Nebraska History* 49, no. 2 (1968): 165–85.

———. *Germans in the New World: Essays in the History of Immigration*. Urbana: University of Illinois Press, 1990.

Marcus, Steven. *The Other Victorians: A Study of Sexuality and Pornography in Mid-Nineteenth-Century England*. New York: Basic Books, Inc., 1966.

May, Henry F. *The End of American Innocence: A Study of the First Years of Our Own Time, 1912–1917*. New York: Alfred A. Knopf, 1959.

Mazlish, Bruce. *The Uncertain Sciences*. New Haven: Yale University Press, 1998.

McClellan, B. Edward. *Moral Education in America: Schools and the Shaping of Character from Colonial Times to the Present*. New York: Teachers College Press, 1999.

McGreevy, John T. *Catholicism and American Freedom: A History*. New York: W. W. Norton and Company, 2003.

McMurry, Sally. "City Parlor, Country Sitting Room: Rural Vernacular Design and the American Parlor, 1840–1900." *Winterhur Portfolio* 20, no. 4 (1985): 261–80.

McWhorter, John. *The Power of Babel: A Natural History of Language*. New York: Perennial, 2003.

Menand, Louis. *The Metaphysical Club.* New York: Farrar, Straus and Giroux, 2001.

Mencken, H. L. *The American Language: An Inquiry into the Development of English in the United States.* 4th ed. New York: Alfred A. Knopf, 1936.

————. *The American Language: An Inquiry into the Development of English in the United States, Supplement I.* New York: Alfred A. Knopf, 1966.

Miaso, Jozef. *The History of the Education of Polish Immigrants in the United States.* Translated by Ludwik Krzyzanowski. New York: Kosciuszko Foundation, 1977.

Miller, Perry. *Errand into the Wilderness.* 2nd ed. Cambridge, MA: Harvard University Press, 1964.

Mintz, Steven. *Huck's Raft: A History of American Childhood.* Cambridge, MA: Belknap Press, 2004.

Mirel, Jeffrey. "Civic Education and Changing Definitions of American Identity, 1900–1950." *Educational Review* 54, no. 2 (2002): 143–152.

————.*The Rise and Fall of an Urban School System: Detroit, 1907–81.* Ann Arbor: University of Michigan Press, 1993.

Nasaw, David. *Children of the City: At Work and at Play.* Garden City, NJ: Anchor Press / Doubleday, 1985.

————. *Schooled to Order: A Social History of Public Schooling in the United States.* Oxford: Oxford University Press, 1981.

Nelson, O. N., ed. *History of the Scandinavians and Successful Scandinavians in the United States.* 2nd ed. 2 vols. Minneapolis: O. N. Nelson and Company, 1900.

Neumann, Richard. *Sixties Legacy: A History of the Public Alternative Schools Movement, 1967–2001.* New York: Peter Lang, 2003.

Newman, Joseph W. "Antebellum School Reform in the Port Cities of the Deep South." In *Southern Cities, Southern Schools: Public Education in the Urban South,* edited by David N. Plank and Rick Ginsberg, 17–35. New York: Greenwood Press, 1990.

Nieto-Phillips, John M. *The Language of Blood: The Making of Spanish-American Identity in New Mexico, 1880s-1930s.* Albuquerque: University of New Mexico Press, 2004.

Norlie, Olaf Morgan. *History of the Norwegian People in America.* Minneapolis: Augsburg Publishing House, 1925.

Oakes, James. "The Ages of Jackson and the Rise of American Democracies." *The Journal of the Historical Society* 6, no. 4 (2006): 491–500.

O'Brien, Jr., Kenneth B. "Education, Americanization and the Supreme Court: The 1920s." *American Quarterly* 13, no. 2 (1961): 161–71.

Olneck, Michael R., and Marvin Lazerson. "The School Achievement of Immigrant Children: 1900–1930." *History of Education Quarterly* 14, no. 4 (1974): 453–82.

Park, Robert E. *The Immigrant Press and Its Control.* Westport, CT: Greenwood Press, 1970.

Perlmann, Joel. *Ethnic Differences: Schooling and Social Structure among the Irish, Italians, Jews, and Blacks in an American City, 1880–1935.* Cambridge: Cambridge University Press, 1988.

Prucha, Francis Paul. *The Churches and the Indian Schools, 1888–1912.* Lincoln: University of Nebraska Press, 1979.

Ramsey, Paul J. "Building a 'Real' University in the Woodlands of Indiana: The Jordan Administration, 1885–1891." *American Educational History Journal* 31, no. 1 (2004): 20–28.

———. "Histories Taking Root: The Contexts and Patterns of Educational Historiography During the Twentieth Century." *American Educational History Journal* 34, no. 2 (2007): 347–63.

———. "'Let Virtue Be Thy Guide, and Truth Thy Beacon-Light:' Moral and Civic Transformation in Indianapolis's Public Schools." In *Civic and Moral Learning in America*, edited by Donald Warren and John J. Patrick, 135–51. New York: Palgrave Macmillan, 2006.

———. "The War against German-American Culture: The Removal of German-Language Instruction from the Indianapolis Schools, 1917–1919." *Indiana Magazine of History* 98, no. 4 (2002): 285–303.

———. "Wrestling with Modernity: Philanthropy and the Children's Aid Society in Progressive-Era New York City." *New York History* 88, no. 2 (2007): 153–74.

Ravitch, Diane. "Politicization and the Schools: The Case of Bilingual Education." *Proceedings of the American Philosophical Society* 129, no. 2 (1985): 121–28.

Reese, William J. *America's Public Schools: From the Common School to "No Child Left Behind."* Baltimore: Johns Hopkins University Press, 2005.

———. *The Origins of the American High School.* New Haven: Yale University Press, 1995.

———. "'Partisans of the Proletariat': The Socialist Working Class and the Milwaukee Schools, 1890–1920." *History of Education Quarterly* 21, no. 1 (1981): 3–50.

———. "The Philosopher-King of St. Louis." In *Curriculum and Consequence: Herbert M. Kliebard and the Promise of Schooling*, edited by Barry M. Franklin, 155–77. New York: Teachers College Press, 2000.

———. *Power and the Promise of School Reform: Grassroots Movements During the Progressive Era.* New York: Teachers College Press, 2002.

———. "Public Education in Nineteenth-Century St. Louis." In *St. Louis in the Century of Henry Shaw: A View Beyond the Garden Wall*, edited by Eric Sandweiss, 167–87. Columbia: University of Missouri Press, 2003.

Reich, Steven A. "The Great Migration and the Literary Imagination." *The Journal of the Historical Society* 9, no. 1 (2009): 87–128.

Reichmann, Eberhard, La Vern J. Rippley, and Jörg Nagler, eds. *Emigration and Settlement Patterns of German Communities in North America.* Indianapolis: Max Kade German-American Center, 1995.

Reuben, Julie A. *The Making of the Modern University: Intellectual Transformation and the Marginalization of Morality.* Chicago: University of Chicago Press, 1996.

Reyhner, John, and Jeanne Eder. *American Indian Education: A History.* Norman: University of Oklahoma Press, 2004.

Riesman, David, in collaboration with Reuel Denney and Nathan Glazer. *The Lonely Crowd: A Study of the Changing American Character*. New Haven: Yale University Press, 1950.

Rippley, La Vern J. *The German-Americans*. Boston: Twayne Publishers, 1976.

Rodriguez, Richard. *Hunger of Memory: The Education of Richard Rodriguez*. New York: Bantam Books, 1983.

Ross, Colin. *Unser Amerika: Der Deutsche Anteil an Den Vereinigten Staaten*. Leipzig: F. A. Brockhaus, 1936.

Rudolph, Frederick. *The American College and University: A History*. 1962. Reprint, with an introductory essay by John R. Thelin, Athens: University of Georgia Press, 1990.

Rury, John L. "Social Capital and the Common Schools." In *Civic and Moral Learning in America*, edited by John J. Patrick, 69–86. New York: Palgrave Macmillan, 2006.

San Miguel, Jr., Guadalupe. *Contested Policy: The Rise and Fall of Federal Bilingual Education in the United States, 1960–2001*. Denton, TX: University of North Texas Press, 2004.

———. *"Let All of Them Take Heed": Mexican Americans and the Campaign for Educational Equality in Texas, 1910–1981*. Austin: University of Texas, 1987.

Sanchez, George I. *Concerning Segregation of Spanish-Speaking Children in the Public Schools*. Austin: University of Texas, 1951.

———. *Forgotten People: A Study of New Mexicans*. Albuquerque: University of New Mexico Press, 1940.

Schlesinger, Arthur M. Jr. *The Age of Jackson*. Boston: Little, Brown and Company, 1946.

Schlossman, Steven L. "Is There an American Tradition of Bilingual Education?: German in the Public Elementary Schools, 1840–1919." *American Journal of Education* 91, no. 2 (1983): 139–86.

Schmidt, Hugo. "A Historical Survey of the Teaching of German in America." In *The Teaching of German: Problems and Methods*, edited by Eberhard Reichmann, 3–7. Philadelphia: National Carl Schurz Association, 1970.

Sealander, Judith. "Curing Evils at Their Source: The Arrival of Scientific Giving." In *Charity, Philanthropy, and Civility in American History*, edited by Lawrence J. Friedman and Mark D. McGarvie, 217–39. Cambridge: Cambridge University Press, 2003.

Sklar, Kathryn Kish. *Catharine Beecher: A Study in American Domesticity*. New York: W. W. Norton and Company, 1976.

Skowronek, Stephen. *Building a New American State: The Expansion of Administrative Capacities, 1877–1920*. Cambridge: Cambridge University Press, 1982.

Smith, Lynn T., and Vernon J. Parenton. "Acculturation among the Louisiana French." *American Journal of Sociology* 44, no. 3 (1938): 355–64.

Smith, Rogers M. *Civic Ideals: Conflicting Visions of Citizenship in U.S. History*. New Haven: Yale University Press, 1997.

Sommerfeld, Linda L. "An Historical Descriptive Study of the Circumstances That Led to the Elimination of German from the Cleveland Schools, 1860–1918." PhD diss., Kent State University, 1986.

Spolsky, Bernard. *American Indian Bilingual Education*: University of New Mexico, 1974.

Stephenson, George M. "Nativism in the Forties and Fifties, with Special Reference to the Mississippi Valley." *Mississippi Valley Historical Review 9*, no. 3 (1922): 185–202.

Storr, Richard. "The Education of History: Some Impressions." *Harvard Educational Review* 31, no. 2 (1961): 124–35.

Struve, Walter. *Germans and Texans: Commerce, Migration, and Culture in the Days of the Lone Star Republic*. Austin: University of Texas Press, 1996.

Takaki, Ronald. *A Different Mirror: A History of Multicultural America*. Boston: Little, Brown and Company, 1993.

———. *Iron Cages: Race and Culture in 19th-Century America*. Revised ed. New York: Oxford University Press, 2000.

———. *Strangers from a Different Shore: A History of Asian Americans*. Boston: Little, Brown and Company, 1989.

Tamura, Eileen H. "The English-Only Effort, the Ant-Japanese Campaign, and Language Acquisition in the Education of Japanese Americans in Hawaii, 1915–40." *History of Education Quarterly* 33, no. 1 (1993): 37–58.

Taylor, A. J. P. *The Origins of the Second World War*. Reissued ed. New York: Touchstone, 1996.

Thompson, E. P. *The Making of the English Working Class*. New York: Vintage Books, 1966.

Todd, Ellen Wiley. *The "New Woman" Revised: Painting and Gender Politics on Fourteenth Street*. Berkley: University of California Press, 1993.

Todd, Lewis Paul. *Wartime Relations of the Federal Government and the Public Schools, 1917–1918*. New York: Teachers College, 1945.

Toth, Carolyn R. *German-English Bilingual Schools in America: The Cincinnati Tradition in Historical Context*. New York: Peter Lang, 1990.

———. "A History of German-English Bilingual Education: The Continuing Cincinnati Tradition." PhD. diss., University of Cincinnati, 1988.

Troen, Selwyn K. *The Public and the Schools: Shaping the St. Louis System, 1830–1920*. Columbia: University of Missouri Press, 1975.

Tyack, David B. *The One Best System: A History of American Urban Education*. Cambridge, MA: Harvard University Press, 1974.

———. *Seeking Common Ground: Public Schools in a Diverse Society*. Cambridge, MA: Harvard University Press, 2003.

Tyack, David B., Robert Lowe, and Elisabeth Hansot. *Public Schools in Hard Times: The Great Depression and Recent Years*. Cambridge, MA: Harvard University Press, 1984.

Vinovskis, Maris A. *The Origins of Public High Schools: A Reexamination of the Beverly High School Controversy*. Madison: University of Wisconsin Press, 1985.

Warren, Donald. "The Wonderful Worlds of the Education of History." *American Educational History Journal* 32, no. 1 (2005): 108–15.

Weber, Max. *The Protestant Ethic and the "Spirit" of Capitalism and Other Writings*. Translated by Peter Baehr and Gordon C. Wells. New York: Penguin Books, 2002.

Weiss, Bernard J., ed. *American Education and the European Immigrant, 1840–1940.* Urbana: University of Illinois Press, 1982.

West, Elliot. *Growing up with the Country: Childhood on the Far Western Frontier.* Albuquerque: University of New Mexico Press, 1989.

Wiebe, Robert H. *The Search for Order, 1877–1920.* New York: Hill and Wang, 1967.

———. *The Segmented Society: An Introduction to the Meaning of America.* New York: Oxford University Press, 1975.

Wilcox, Clifford. "World War I and the Attack on Professors of German at the University of Michigan." *History of Education Quarterly* 33, no. 1 (1993): 59–84.

Wilentz, Sean. "Politics, Irony, and the Rise of American Democracy." *The Journal of the Historical Society* 6, no. 4 (2006): 537–53.

Wiley, Terrence G. "Heinz Kloss Revisited: National Socialist Ideologue or Champion of Language-Minority Rights?" *International Journal of the Sociology of Language* 154 (2002): 83–97.

Wilson, James. *The Earth Shall Weep: A History of Native America.* New York: Grove Press, 1998.

Wittke, Carl. *The German-Language Press in America.* Lexington: University of Kentucky Press, 1957.

———. *Refugees of Revolution: The German Forty-Eighters in America.* Philadelphia: University of Pennsylvania Press, 1952.

Wood, Gordon S. "The Localization of Authority in the 17th-Century English Colonies." *Historically Speaking* 8, no. 6 (2007): 2–5.

———. *The Radicalism of the American Revolution.* New York: Vintage Books, 1993.

Wraga, William G. "A Progressive Legacy Squandered: The *Cardinal Principles* Report Reconsidered." *History of Education Quarterly* 41, no. 4 (2001): 494–519.

Zamora, Emilio, Cynthia Orozco, and Rodolfo Rocha, eds. *Mexican Americans in Texas History: Selected Essays.* Austin: Texas State Historical Association, 2000.

Zeydel, Edwin. "The Teaching of German in the United States from the Colonial Times to the Present." *The German Quarterly* 37 (1964): 315–92.

Ziegler, James P. *The German-Language Press in Indiana: A Bibliography.* Indianapolis: Max Kade German-American Center and Indiana German Heritage Society, Inc., 1994.

Zimmerman, Jonathan. "Ethnics against Ethnicity: European Immigrants and Foreign Language Instruction, 1890–1940." *Journal of American History* 88, no. 4 (2002): 1383–404.

Zinn, Howard. *A People's History of the United States, 1492–Present.* New York: Perennial Classics, 2001.

Index

Note: Numbers in bold indicate tables